Engineering Design Modeling

with

Unigraphics®

Robert A. Chalou
Michigan State University

KENDALL/HUNT PUBLISHING COMPANY
4050 Westmark Drive Dubuque, Iowa 52002

Certain copyrighted materials are used in the textbooks under a license from Unigraphics Solutions Inc. (UGS). However, this textbook has not been approved for use in conjunction with Unigraphics software by UGS, and is not intended to be considered UGS documentation. Accordingly, UGS disclaims all responsibility for the information contained in this textbook or any errors which may occur as result of its use.

ISBN 0-7872-7757-6

Printed in the United States of America
10 9 8 7 6 5 4 3 2 1

This is dedicated to my wife Cindy
and our children Becky and Carl.

Engineering Design Modeling

Table of Contents

Sketcher

Solid Modeling

Foreword

The purpose of the book is to give students a sound, basic knowledge of 3-D Solid Modeling and Assemblies using Unigraphics®. This book is by no means a complete operator's manual for the software. Instead, it is intended to give students the tools to explore the software and the courage to try new functionality, as they become more versatile designers. Students are encouraged to learn 3-D Solid Modeling early in their education. Learning it early tends to give students a better ability to experiment with shape, dimensions and design intent. Thereby, they are much freer to be creative and concentrate on the design of the product rather than the final dimensions.

All part files listed in the book are available for downloading at http://www.egr.msu.edu/classes/msm160/parts. There are folders for each chapter.

I would like to especially thank Hulas King of Unigraphics Solutions™, for supplying me with the motivation and support to accomplish this task and Unigraphics Solutions™ for supplying me with all screen captures used in this book. Also, thanks to William Hehs, Donald Papiez and Dr. Vassilios Theodoractos of General Motors Knowledge Center for helping to define the direction that this book should take. Thanks also to Dr. Vassilios Theodoracatos for providing the puzzle graphic, depicting the idea of Unigraphics integration within the engineering curriculum, used on the cover. Special thanks to CAD Potential, Inc., for providing an excellent source of reference material via their web-based UG training courses.

Chapter 1

Basic UNIX

The Common Desktop Environment:

CDE is a windowing environment common to UNIX-based workstations. Most of the features mentioned in this chapter are common to all systems using CDE. Because features can easily be added to or removed from the operating environment, functions common to your location may not be detailed. This chapter concentrates on functionality that is basic and generic to virtually all locations.

Note: The appearance of CDE screen varies slightly from one hardware plaform to another. The screen examples in this manual are based on the Sun and Windows NT platforms.

The Unigraphics Workstation:

A Unigraphics workstation is comprised of several different pieces of hardware. The standard equipment is a processing unit, monitor, keyboard, and a mouse.

Logging in to the server:

Before you can access the CDE environment, you must first login to the network. The systems administrators will assign you a user ID, and you will practice logging-in to the file server network to verify that your user ID is working correctly.

Log in Procedure:

In the user name box, type in your user ID and click OK (or press the RETURN key not the Tab key).
Note: Capital letters and spaces are not allowed as part of login ID. Please check the options pull down menu to make sure that the session is set for Common Desktop Environment.
Make sure that you are entering a CDE session. At the login screen check Options | Session | Common Desktop Environment
In the password box, type in your password and click OK (or press the RETURN key).

Note: Your password will not be visible as you enter it.

The Desktop and Workspaces:

The first thing you see after you log in is the Desktop and the Front Panel. The Front Panel is typically located at the bottom of the desktop. This is where most of the programs, or applications are started.

The background of a graphical interface is typically called the desktop or workspace. It is the area of the monitor screen where you do work or display objects. In CDE, you can have multiple desktops (workspaces) operating at one time.

Workstation Front Panel:

The front panel is used to collect icons and is located on the desktop. These icons are used to launch applications within the CDE. Placing the cursor on an up arrow and clicking opens the sub-panels. Clicking on the down arrow closes them.

These are the icons and programs commonly attached to the Front Panel.

Calendar  Style Manager  Mail  File Manager

 Terminal Window  Printer  Help  Trash Can

The following options are specific to the Front Panel.

Workspace Buttons:

A CDE has at least one active workspace but is capable of many. Multi-tasking incorporates the use of several workspaces; each workspace can be dedicated for specific programs. This will reduce the amount of information in a single workspace and allow quick access to other open programs.

The active workspace is the one that appears to be depressed. Placing the cursor over the desired workspace button and right clicking will pop-up a menu allowing for the workspace to be customized. A workspace may also be renamed by double clicking on the desired workspace button, entering a new name, and then pressing RETURN. Renaming of workspaces is a common practice. Most of the time the names found on the workspace buttons are the names of the programs used in that particular workspace. It is a good idea to rename one of the desktops for Unigraphics as a dedicated workspace.

Clock:

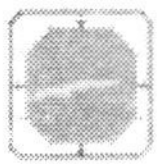 This icon is used to display the current time set by the processing unit.

Exit:

 This is used to exit the session. A session may also be exited by placing the cursor on the backdrop, right clicking and selecting Log Out.

Busy Indicator:

 This icon will flash to indicate that the processor is busy.

Style Manager:

 The Style Manager is used to set the preferences for the CDE. Selecting the Style Manager icon from the Front Panel activates the Style Manager window.

Color:

 This is used to specify the color scheme used in the active workspace.

Font:

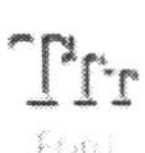 This changes the default font size for CDE applications.

Size: This listing is used to highlight the desired font size.

Backdrop:

 This is used to change the backdrop, or "wallpaper," in the current workspace. Each workspace can have a different backdrop to help differentiate between them.

Mouse:

This is to switch between right and left handed configurations, set the double-click rate; control the acceleration of the mouse, and to define the threshold.

Screen:

This is used to control the appearance of the screen during periods of inactivity.

Auto Screen Blanking:
This is used to enable/disable auto blanking. If turned on, the screen will blank after the set number of minutes without movement from the mouse.

Window:

This is used to control various aspects of window behavior.

Point In Window To Make Active:
This makes a window active when the cursor is over the window.

Click In Window To Make Active:
This makes a window active by clicking when the cursor is in the window.

Raise Window When Made Active:
This raises the active window to the top of the stack of all opened windows in the desktop.

Allow Primary Windows On Top:
This keeps an active window totally visible in the workspace.

Place On Workspace:
This places all minimized windows on the backdrop of the workspace.

Startup:

This is used to control the system behavior during the logging in and out process.

Resume Current Session:
This is used to restart all applications that were running the next time logging in.

Return to Home Session:
This is used to start a default set of applications when logging in.

Logout Confirmation Dialog:
This is used to enable/disable the confirmation dialog when logging out.

Set Home Session:
This is used to define the default set of applications that make up the home session.

Mail:

 This program works similar to any standard e-mail system.

File Manager:

 The File Manager is used to maintain part files created by the applications available in the CDE.

The icon next to the name indicates the file type. A directory always looks like a folder. Typically, text and part file icons look like a piece of paper with writing. Most graphics files have an icon with a picture.

The Directory Structure:

Data is saved to files; the files then are organized into directories. Directories may also contain other directories; called subdirectories. Each user will have a special directory to store their class work in.

Directories:

When opening, closing, and saving a part file it is important to know where the file is located. Directories can be thought of as "folders" that contain information that should be grouped together.

Most computer systems contain many different directories and many levels of directories. A directory may be nested within another directory. This is known as a subdirectory. A directory that contains the directory you are currently in is known as the parent directory.

Subdirectories are used to keep files organized. Directories take very little additional storage space and are easily created.

File Types:

The file that contains the data that is created when designing in Unigraphics is called a part file. This file has a ". prt" file extension.

Each file in a directory has a unique name, but there may be hundreds or thousands of files on any given workstation. The way that a particular file is identified out of all the other files is with its path. For example, let's say that you have a file called "first.txt" in your subdirectory "username". This file's path name would be "/home/u_/username/first.txt".

Note: UNIX does not allow spaces in file names. Problems are avoided by using combinations of letters, numbers, periods, and underscores (_) for file and directory names. In addition, UNIX is case sensitive, so the file "First.txt" is considered to be different than "first.txt."

Changing Directories:

Each File Manager window shows only one directory at a time. There are several ways to change the directory being viewed. The first is to click on the file path line to activate it and then type in the name of the directory. A backslash is used to separate each level of directories. The second is to double-click on a directory icon in the window. The Go Up icon can also be double-clicked to access the directory above the currently displayed directory.

Moving Files:

Files can be easily moved between two opened File Manager windows. You must have write access to the directory or the move will fail. This is accomplshed by dragging the file icon to the second File Manager window.

To move a file to a subdirectory of the current directory, a second window is not needed. Drag the file icon over the folder icon of the directory where it is to be moved.

Creating and Editing in a Directory:

With File Manager You Can:

- Create a subdirectory in the current directory.
- Rename a file or directory.
- Copy a file with the drag and drop technique.
- Delete a file by dragging the file icon into the Trash Can.
- Change permissions assigned to a file or a directory. The permissions can be modified to restrict users from making changes to part files. Files and

directories have Read/Write/Execute settings for the Owner, Group, or Other. Owner refers to the person having custody of the file. Read access allows someone to open a file. Write access allows someone to modify a file.

Exercise: Creating a new Directory

1. Select the File Manager icon.
2. Select **File | New** Folder.
3. Type in test for a name for the directory folder and press the RETURN key.
4. Create a file in the folder named **test_1**.
5. Create another file named **test_1a**.

Exercise: Renaming a file or Directory

1. Click on the file or directory name.
2. A text field appears around the current name.
3. Type the new file or directory name.
4. Press the RETURN key.

Exercise: Copying a file with drop and drag

1. Select a file.
2. Hold the CTRL key while dragging the file to the directory where it is to be copied.

Exercise: Deleting a file by dragging

1. Select one of the files.
2. Drag the file icon into the Trash Can.

Exercise: Deleting a file with the pop-up menu

1. Select a file or directory icon by clicking on it.
2. Select Put In Trash from the File pop-up menu.

Exercise: Changing file and directory permissions

1. Right click on the file or folder that you want to change.
2. Select Properties from the pull-down menu. The Permissions dialog box appears.
3. Select the appropriate check boxes for Permissions. Change the file to Read/Execute for Owner and Group.
4. Click **OK**.

Note: Unigraphics part files should never be renamed with the File Manager because all links to the file will be lost.

Terminal Window:

 The Terminal Window is used to perform program operations and maintenance through the use of UNIX commands.

Printer:

 The Printer program is used to display the printer queue manager and act as a drop zone for items to be sent to a printer or plotter.

Help:

 The Help program is used to obtain information on the CDE interface.

Trash Can:

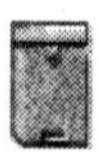 Stores all deleted files until they are shredded. Files may be shredded or restored from the trashcan.

Exercise: Shredding items in the trash can

1. Select the Trash Can icon from the Front Panel.
2. **File | Select All**.
3. **File | Shred**.
4. Click **OK**.
5. **File | Close**.

Exercise: Restore items from the trash can

1. Select the Trash Can icon from the Front Panel.
2. **File | Select All**.
3. **File | Put Back**.
4. **File | Close**.

Chapter 2

Getting Started with Unigraphics and Menus

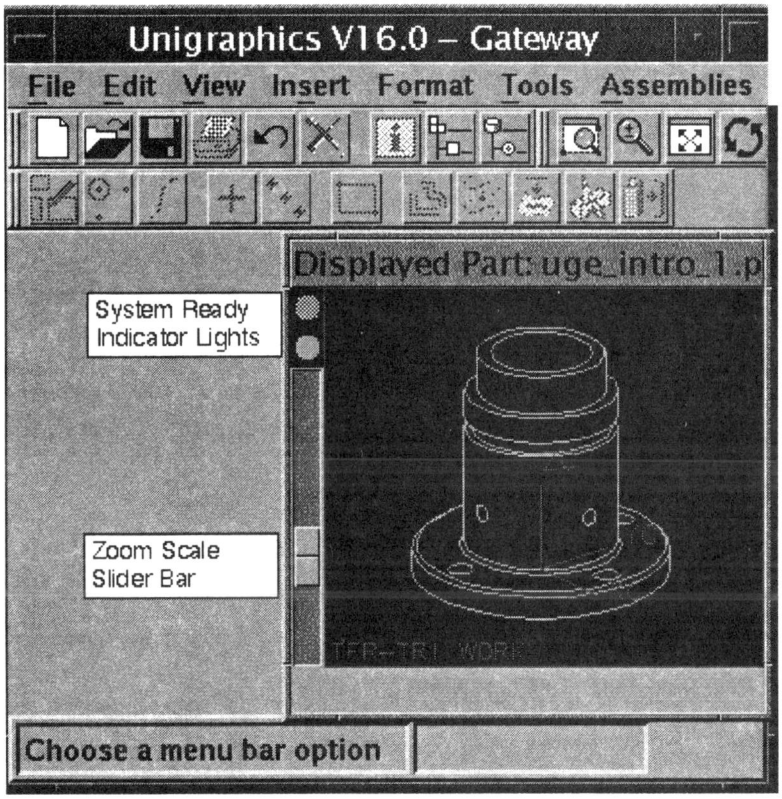

After completing this chapter you will be able to:

- Start Unigraphics.
- Create a New Part File.
- Open an Existing Part File.
- Save a Part File.
- Change the Displayed Part.
- Close a Part File.
- Reopen a Part File.
- Exit Unigraphics.
- Use the View Pop-Up Menu.
- Rotate a View.
- Orient a View.

- Select Objects.
- Blank and Delete Objects.
- Edit the Object Display.
- Create Layer Settings.
- Move Geometry to New Layers.
- Understand the Role of the Absolute Coordinate and Work Coordinate Systems.
- Relocate the Origin of the WCS.
- Rotate the WCS.
- Reorient the WCS in Space.
- Control the Display of Objects.

Starting Unigraphics:

If using UNIX, the Uniproducts activity window appears. From here you can specify the application needed, (in the beginning we will use the default Unigraphics). With NT you will choose the Icon on your desktop or through **Start | Programs | Unigraphics V (your version) | Unigraphics**.

The UNIX command for starting Unigraphics is: **ugmenu** at the command prompt in a terminal window.

Start a Unigraphics Session.

Main Window Features:

After a Unigraphics session is activated the Unigraphics main window appears. The Unigraphics environment consists of multiple windows. The main menu bar is at the top and the Cue line and Status line are initially at the bottom. Other windows overlay this window. The dialog box area for application dialog boxes is to the left of the Graphics window.

Open a part file. Your instructor will supply you with a file.

Main Menu Bar:

The main menu bar displays the name of the menu containing related commands. Shaded commands are currently unavailable.

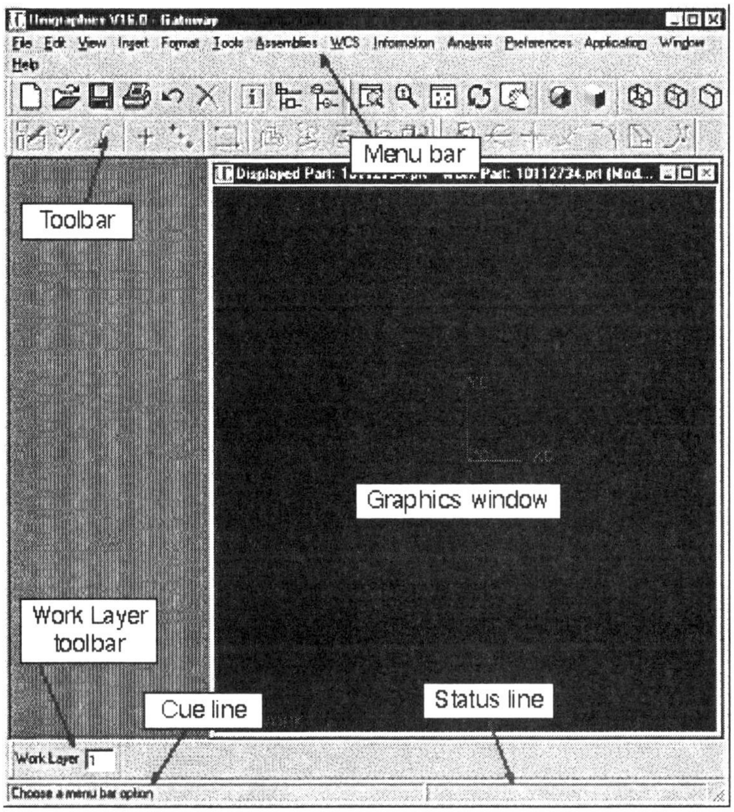
Menu bar
Toolbar
Graphics window
Work Layer toolbar
Cue line
Status line
Work Layer
Choose a menu bar option

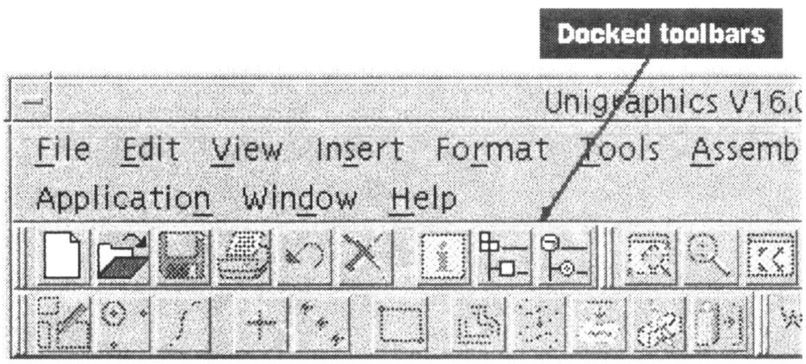
Docked toolbars
File Edit View Insert Format Tools
Application Window Help

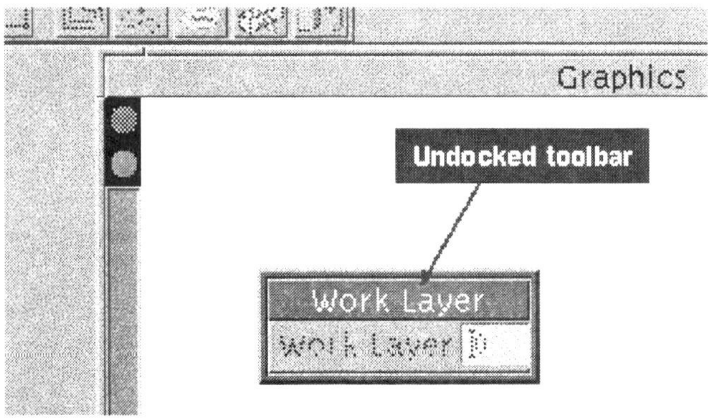
Graphics
Undocked toolbar
Work Layer
Work Layer

The Standard Tool Bar:

The standard toolbar is a user tool that automatically loads when Unigraphics starts. It consists of a row of icons representing commonly used options such as **File | New** and **File | Save**.

Cue Line:

The Cue line is located either just below the tool bar in the Unigraphics window or below the Graphics window. The position of the Cue line is controlled by going to **Tools | Customize | Options**, and selecting the position. It displays important information indicating the next course of action and prompts you for next selections as necessary.

Status Line:

The Status line is located to the right of the Cue line. The Status line provides feedback about the current option or about the just completed operation.

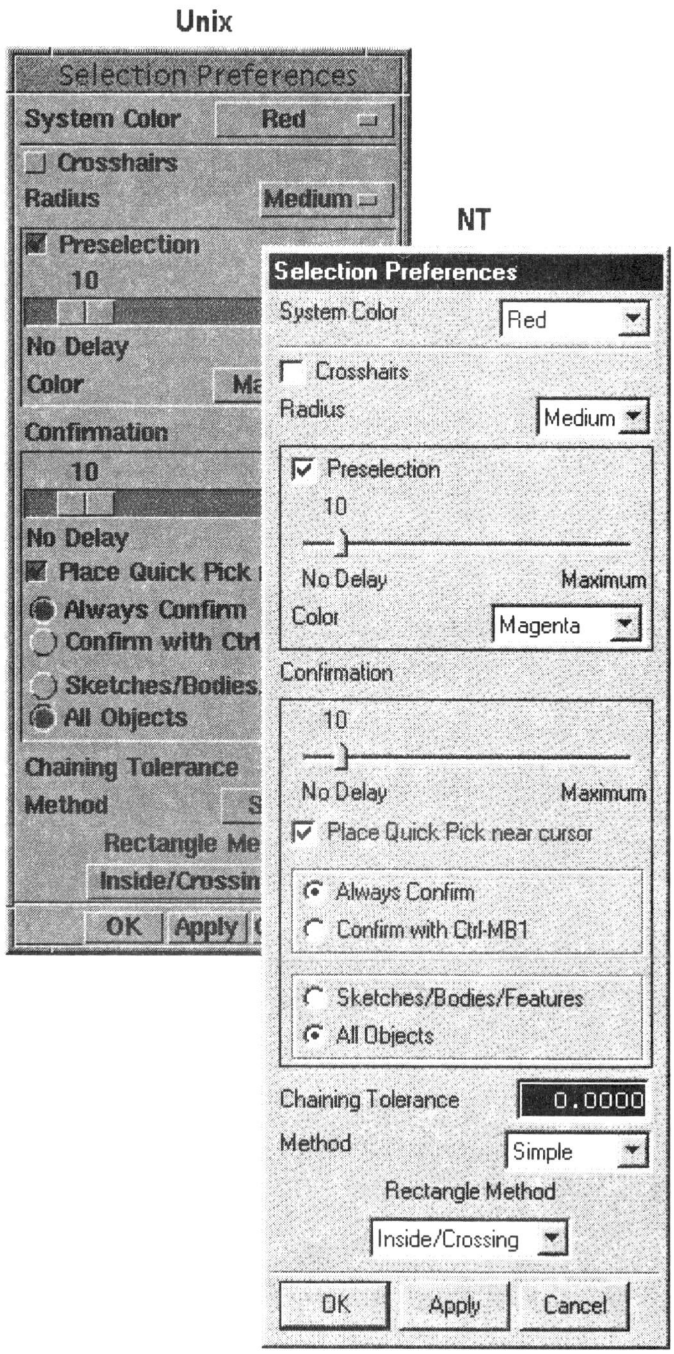

Graphics Window:

This is the window in which the Unigraphics part file displays.

Crosshairs and the Selection Ball:

The selection ball is the round circle surrounding the crosshairs whenthe cursor is in the Graphics window. Objects must be within the selection ball in order to be selected.

Work Part:

The Work Part is the part you are currently working on (since it is possible to have more than one part loaded). There may also be text following the Work Part name indicating that the Work Part is **Read Only** or that the Work Part is **Modified**.

Displayed Part:

This is the part currently displayed in the Graphics window. Unigraphics displays the Displayed Part name in the header of the Graphics window.

Work Layer:

The Work Layer dialog box, displays the layer currently active in the Graphics window. You can change the Work Layer by typing the desired layer number in the field.

Messages:

Message windows are displayed after various commandactions are performed in Unigraphics. These message windows provide information regarding the action either before or after the command is executed.

Information Message:

An information message conveys important information about a particular task. Clicking **OK** continues the operation.

Warning Message:

A warning message displays information that warns of results that you cannot reverse.

Error Message:

Error messages display at the current cursor location. To continue, the message must be acknowledged by clicking **OK**. If the message repeats, press the ESC key to close the window.

Question Dialog:

A question dialog displays a question you must answer before Unigraphics continues its operation. Picking **OK** continues the operation. **Cancel** cancels the operation.

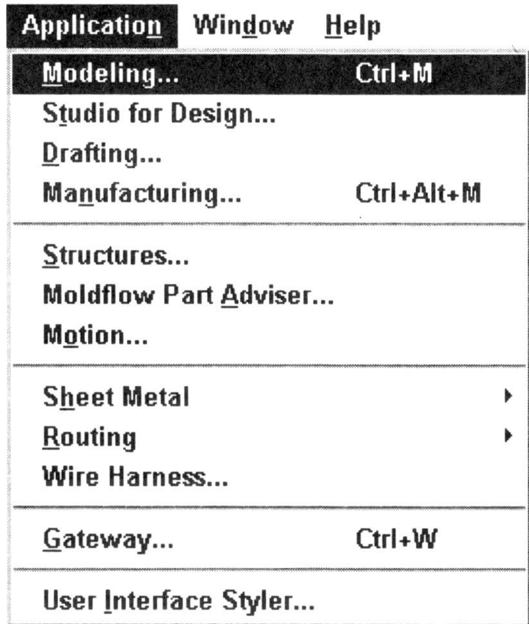

Unigraphics Applications:

Unigraphics is made up of separate applications. Each application provides bols to perform different job tasks within Unigraphics. When **Application** is selected in Unigraphics, the various applications list as shown in the menu.

Gateway:

Gateway is the fundamental application that all other applications depend on. This is also the application running when a UG session is opened. Gateway defines the user interface and allows you to perform some fundamental operations such as viewing geometry, obtaining information about the geometry, and plotting drawings. It is also literally a gateway into all of the other applications.

Modeling:

This provides Modeling applications such as feature based modeling and sketcher.

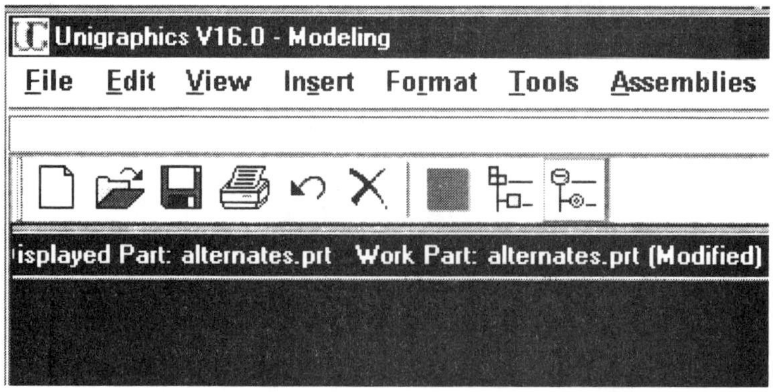

Toolbars:

Toolbars allow you to customize your personal workspace. You can define which toolbars you wish to display and locate them anywhere on your desktop, either docked to the Unigraphics window, or floating.

Docking Toolbars:

Initially a toolbar displays as a floating window. It can be relocated by dragging it to a new location, or docked by dragging it to the Unigraphics window. The toolbar may be docked along any of the four sides of the UG window.

Multiple toolbars can be docked next to each other and are separated by "grip handles."

Customizing Toolbars:

The toolbars can be displayed by using **View** | **Toolbars**. Click to display a toolbar; a check mark indicates a toolbar is displayed.

Toolbars can be customized using **Tools** | **Customize**. The four top buttons, such as **Toolbars** and **Commands**, change the display of the options in the rest of the dialog. Using the **Customize** dialog, you can define which toolbars and commands display, as well as the icon size.

Note: On UNIX operating systems, you can only dock the toolbars at the top of the Unigraphics window.
Customize some of your toolbars and remove the items you rarely use. This keeps your desktop clean and reduces the time spent searching for icons you use regularly

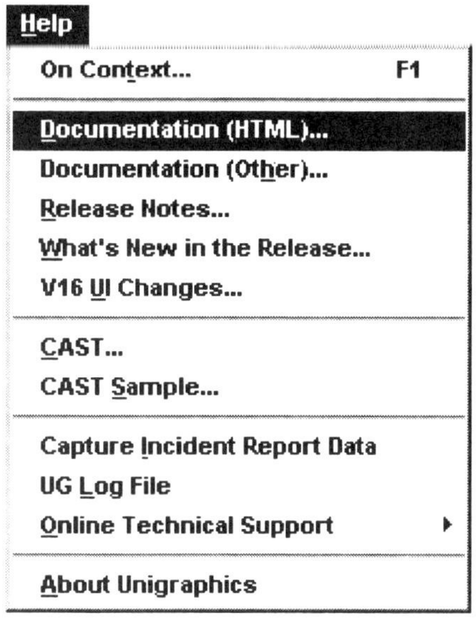

Unigraphics Help Options:

The **Help** menu contains options to access online documentation covering all aspectsof Unigraphics.

Documentation:

This command opens a Web browser window with Unigraphics Solutions' documentation set. Selecting **Documentation** from the Help menu opens the current version of Unigraphics documentation. The documentation supplies instructions for specific operations, which you can apply as needed. The documentation is always available, and is a great resource for new tools and techniques to add to your Unigraphics skills.

Searching for Information:

A quicker way to find something in the table of contents is to use **Search** in the on-line documentation window menu bar.

The **Search Text** dialog is displayed after selecting **Search** | **Text** from the menu. Here a word or topic is entered in the text box. This word can be searched forwards or backwards through the active document. If the word or phase is located, it is highlighted in the document.

CAST:

This runs Unigraphics Solutions' computer assisted self teach.

About Unigraphics:

This displays information about Unigraphics Solutions Inc., andyour system

Creating a New Part File:

File | New

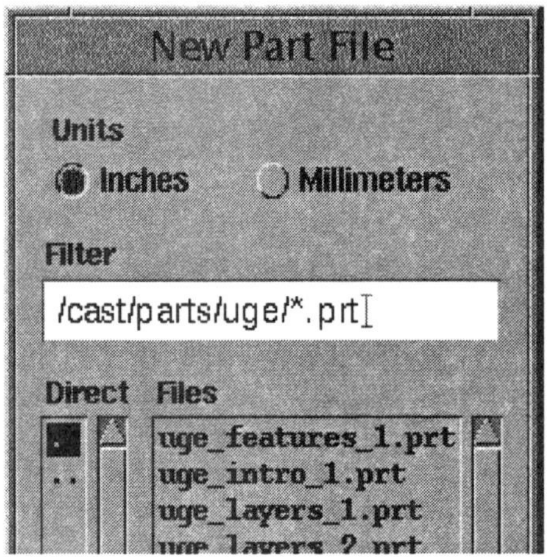

The file exists temporarily in the computer's memory until you save the file. Saving the file adds it to the directory folder. It is listed by name in the folder.

File Name:

Input the desired name for your file in this text box.

File of Types:

This list allows you to select the kind of file, such as **.prt**; you need to create in the selected directory. Only UG part files can be created in this dialog.

Units:

This group box lets you select the base unit of measure for the part file When values are entered, they are interpreted in the base units. For example, if the units are **Millimeters**, entering a length of "50" implies 50 millimeters.

Exercise: Creating a new part file

1. Select **File | New**.
2. Select **inches** as the type of measurement **Units**.

3. Navigate to the directory you want to store the file in, (your instructor will inform you of the proper directory).
4. Input **seed_in** as a new file name in the **File name** field.
5. Click **OK** to create the new file.
6. Also create a millimeter seed file and name it **seed_mm**.

Opening as Existing Part File:

File | Open

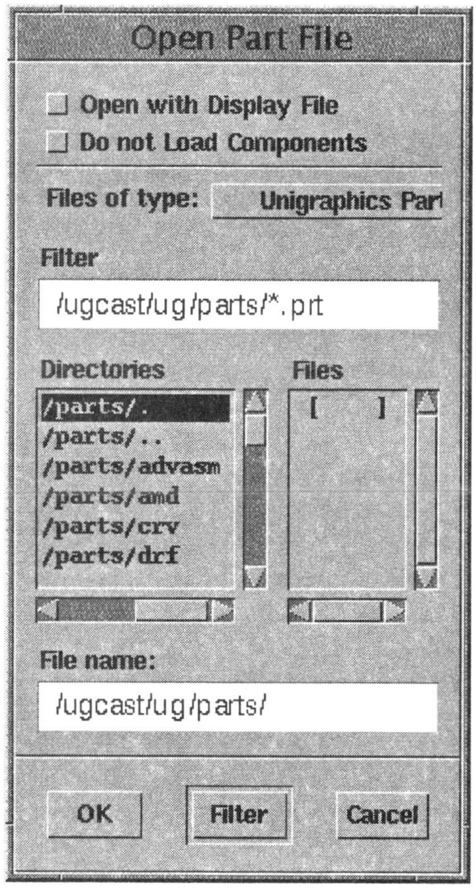

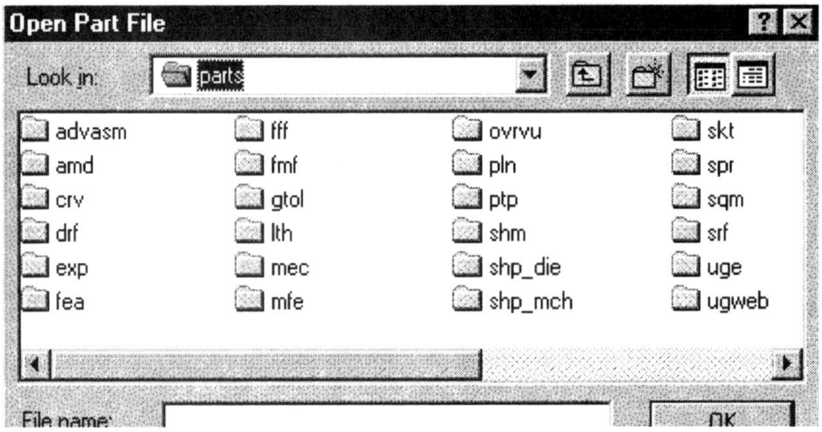

Look in (for NT) and Filter (for UNIX).

File Name:

Use this field to enter the name of the desired file.

Files of Type:

When you open a file, you can select from four different file types. This is used to filter out part files by their file extension.

.prt
This is used for a Unigraphics part file.
.bkm
This is used for a Unigraphics bookmark file.

Do Not Load components: This prevents the components of an assembly from loading when the assembly is opened.

Exercise: Open a part file

1. Select **File | Open**.
2. Use the **Look in** list to navigate to the directory containing the files.
3. Select, **.prt** as the type of file to open from the **Files of type**.
4. From the file list, select **menus_1,** you will work with this file for the rest of this chapter.
5. Click **OK**.

Note: If you attempt to open a part file that is already open, but not currently displayed, a dialog appears allowing that file to be displayed or reopened, or lets you cancel the opening process altogether.

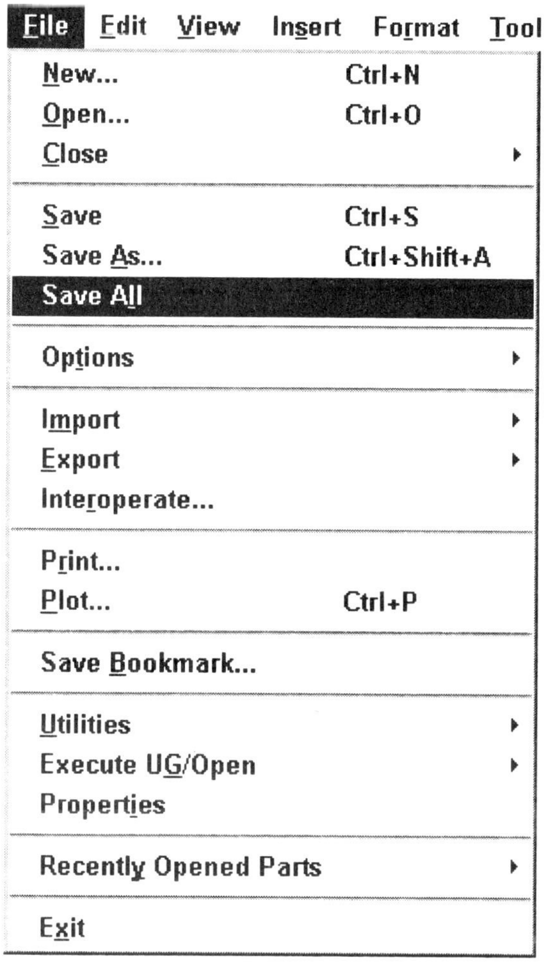

Saving Part Files:

File | Save

Save:

This saves the work part. After the part file is saved, the "Modified" message at the top of the Graphics window disappears. As soon a modfication takes place, the "Modified" message once again displays.

Save As:

This saves the work part using a different name. When **File | Save As** is selected, the **Save Part File As** dialog displays.

Save All:

This saves all the part files currently open. After the parts are saved, a "Saving all parts succeeded" message displays in the Status line.

Save and Close:

Use this to save and then close the work part.

Save As and Close:

Use this to save the work part under a different file name and then closethe file.

Save All and Close:

Use this to save and then close all open part files.

Save All and Exit:

Use this to save all open part files and then exit Unigraphics.

For the following exercises you must reopen the file the instructor supplied earlier.

Exercise: Saving a part file (you must have write permissions to save a file)

1. Select **File** | **Save**.

Exercise: Save the file using a different name or directory

1. Select **File** | **Save As**.
2. Navigate to a directory to save the part file in.
3. Input the new name in the **File name** text box.
4. Click **OK**.

Exercise: Save the part and close the file

1. Select File | Close | Save and Close.

Closing the Part File:

File | Close

Close Selected Parts:

This closes selected part files that are open. When this is selected, the **Close Part** dialog displays.

The file list displays all the part files currently open. Select the part file to close by selecting it from the list. Select multiple part files by pressing the **CTRL** key when selecting the files.

File Filter List

Selecting either **Top Level Assembly Parts** or **All Parts in Session** changes the listing of the selectable files. **Top Level Assembly Parts** only lists assembly files, while **All Parts in Session** lists all part files open in the session.
Close all

Selecting this closes all parts in the list. It is identical to the **File | Close | All Parts**.

Close All Parts:

This closes all the open parts. If a modified part file is not saved, a **Question** dialog displays. Select **OK** to close the part files without saving them. Select **Cancel** to cancel the Close operation.

Views:

A view is a depiction of the part file from a certain perspective.

It is important to note that the part is fixed in CAD-space. You can imagine you are flying around the part and looking at it from different angles. You can fly directly above the part and look down at the top of the part. You can fly below the part and look up at the bottom of the part. You can also fly to view the part from any other viewpoint imaginable. This is called the view "orientation."

Unigraphics provides the following eight commonly used views, called "default" views: **Top**, **Front**, **Right**, **Back**, **Bottom**, **Left**, **TFR-TRI**, and **TFR-ISO**. Each view has its own display scale, center, and borders. These views cannot be deleted.

TFR-ISO:

TFR stands for TOP-FRONT-RIGHT. ***ISO*** is short for "isometric," which means "same measure." In the TFR-ISO view, the projections of the axes of the absolute coordinate system on the flat screen all have the same length. If you view a cube from the TFR-ISO view, you look from the top, front- right corner, directly down the diagonal to the back, bottom-left corner. The display of the cube on the flat screen appears to be a regular hexagon.

TFR-TRI:

The TFR-TRI view is also from the TOP-FRONT-RIGHT direction but the projections of the axes of the absolute coordinate system are not equal. *TRI* is short for "trimetric," which means "three measures." In the TFR-TRI view, the projection of the Y-axis of the absolute coordinate system on the view plane is 7/8 as long as the projection of the Z-axis and the projection of the X-axis is 3/4 as long as the projection of the Z-axis.

Refresh:

This redraws the graphics display. Use it if objects are removed from the Graphics window and other objects are uncovered. You can also use it to remove a temporary display mark such as the asterisk markers left by the Point Subfunction.

Fit:

This changes the view scale so that all displayable objects are within the clipping bounds. Geometry that is blanked or on an invisible layer is not considered.

Zoom:

This adjusts the display scale of the graphics by magnifying the work view. The cursor is displayed in the Graphics window as a magnifying glass. To define a new viewing area, click and drag a rectangle in the Graphics window. Release MB1 and the area inside the rectangle will expand to fill the window. Right-click to discontinue the zoom function.

Zoom In/Out:

This command allows you to use the mouse to zoom in and out without performing two separate operations, as in the **Zoom** command. Position the magnifying glass over the point you want to magnify, pressing MB1 move the cursor up to zoom out and down to zoom in. MB2-click cancels the zoom functionality.

Rotate:

This rotates objects in the work view to a new orientation. Instead of a fixed axis, the axis of rotation is defined by the cursor as you click and move the mouse. Horizontal linear movement across the screen rotates the part around the vertical display axis. Vertical linear movement rotates the part around the horizontal display axis. The cursor becomes a double-arrow rotation symbol when this is activated. MB2-click cancels the dynamic rotation.

Pan:

This allows you to move the view on the plane of the Graphics window. To pan an area in the work view, place the cursor in the Graphics window, right-click and move the cursor down to the **Pan** option to select it. Press MB1 and move the cursor in the direction you want to move. MB2-click cancels the **Pan** function.

Update Display:

This regenerates the geometry in the Graphics window.

Restore:

This restores the view of the model to the orientation it was in prior to the application of a **Zoom**, **Rotate**, or **Pan** operation.

Display Mode:

There are five **Display Mode** options to choose from, but for this course we will discuss three:

Wireframe displays a solid edge representation of the part.

Fully Shaded displays a dynamic shaded image of the body, and is only available when using the 3D driver.

Partially Shaded displays a partially dynamic shaded image of the body, and is only available when using the 3D driver.

Hidden Edges:

Hidden edges are solid edges you should not be able to see because they are behind solid faces. This is used to change the hidden line display in the work view. When this is selected, the following options are available:

Visible displays all edges in a solid line font.

Gray Thin displays all edges in a solid line font and hidden edges in a thin gray font

Dashed displays all edges in a solid line font and hidden edges in a dashed line font.

Invisible displays all edges in a solid line font and hidden edges are invisible. When you rotate the part, the hidden lines do not automatically update. To regenerate the hidden lines, you must select **Update Display**.

Orient View:

This changes the orientation of the current work view to one of the eight canned views. This does not replace the work view. It only changes the viewing direction of that view.

Replace View:

This replaces a view with one of the canned views (Top, Front, Right, Back, Bottom, Left, TFR-ISO, TFR-TRI).

Undo:

This steps the file back to its condition preceding the last executed command.

Note: Use **Orient View** when you want to use your current view visualization preferences (shading, hidden line display) to see the part in another orientation. Using **Replace View** changes the view visualization preferences to those saved for that specific view.

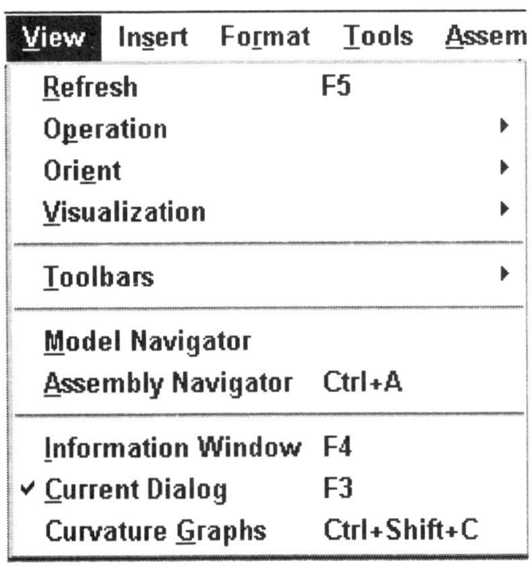

View Menu Options:

The **View** menu contains many of the same options found in the **View** pop-up menu. Some options are similar but contain additional functionality.

Refresh:

This redraws the graphics display. Use this if objects are removed from the Graphics window and other objects are uncovered or to remove a temporary display mark such as the asterisk markers left by the Point Subfunction.

Orient:

This submenu is used to change the orientation of the view based on system coordinates or part geometry.

Toolbars:

This submenu allows you to select which toolbars display. You can also access the **Customize** options from this menu.

Model Navigator, (MNT):

This displays or hides the Model Navigator.

Assembly Navigator, (ANT):

This displays or hides the Assembly Navigator.

Information Window:

This displays or hides an Information window.

Rotate:

This is used to rotate the view. It contains different functionality than the **Rotate** located on the View pop-up menu.

Delete:

This allows a user-defined view(s) to be deleted by selecting from a list of the available views. Views currently displayed on the layout and Unigraphics default views cannot be deleted.

Editing Operations:

Open a part supplied by the instructor. This menu provides options to undo delete or edit objects in the part file. The **Edit** menu changes, depending on the active application.

Note: Use **Undo List** to step back sequentially through the previous operations, restoring the geometry to the state it was in prior to the operation. This list can contain an unlimited number of operations, but is cleared when the part file is saved or when **Application** menu selections are used.

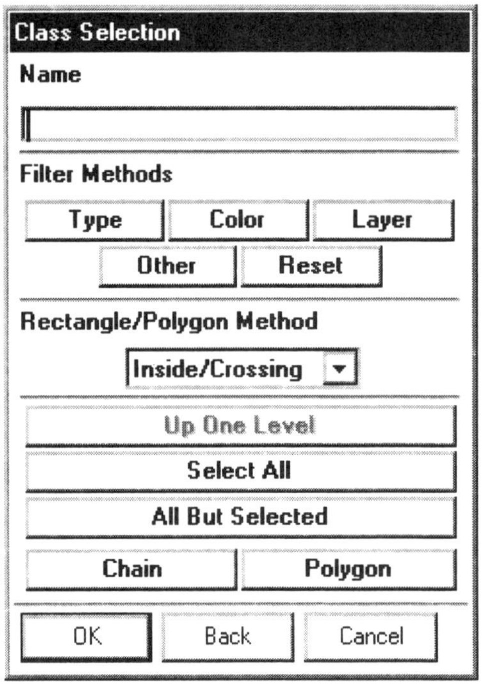

Object Selection:

There are several methods available for selecting objects in Unigraphics. You can select objects directly off the Graphics window, or use the **Class Selection** dialog to help group objects together. There are also options available to assist in selecting geometry hidden by other geometry. Items can also be deselected if necessary.

Selecting Objects:

To select an object, move the selection ball over the object in the Graphics window and pick the object. You can also use the **Class Selection** dialog to help filter the objects available for selection. In most operations, when you need to pick an object, the **Class Selection** dialog automatically displays.

You have the option to activate a pre selection feature that affects the object selection process. This automatically highlights the object within the selection ball. The object temporarily turns a different color, designating it as the current object for selection. The Preselection Color option located on the Selection Preferences dialogdetermines the color.

When selecting an arc or ellipse, Unigraphics highlights the center of the arc or ellipse of the object. As you move the selection ball over the arc or ellipse center, a temporary marker is displayed indicating the center of the object Unigraphics will choose.

In cases where multiple curves, edges or faces are located within the selection ball radius, Unigraphics displays a Quick Pick menu allowing you to cycle quickly through the potential objects. When the cursor is over the object number displaying in the **Quick Pick** dialog, the object highlights.

Deselecting Objects:

To deselect objects, press SHIFT and use the single select method to pick an already highlighted object. The object is deselected. This method of pressing the SHIFT key also works with the Rectangle and Polygon methods.

To deselect the last object and select a new object at the same time, press the SHIFT and CTRL keys at the same time and select the new object.

The **Back** option in the **Class Selection** dialog deselects all objects and closes the dialog.

Using the Class Selection dialog:

This selects objects in the Graphics window. Filter the picked objects by color, type, layer or other specific attributes.

Exercise: Select objects by type

1. Perform an operation such as **Delete**, **Transform**, or **Blank** to display the **Class Selection** dialog.
2. Select **Type** to display the **Select by Type** dialog. All the available object types highlight.
3. Pick the object types you want available for selection.
4. Press CTRL to make multiple selections.
5. Select **OK** to redisplay the **Class Selection** dialog.
6. Pick objects in the Graphics window using any of the selection methods. Only the object types highlighted in the **Select by Type** dialog are available for selection. Select **OK** to perform the operation.

Note: You can have Unigraphics find the end of the chain by selecting **OK** or **End Chain**.

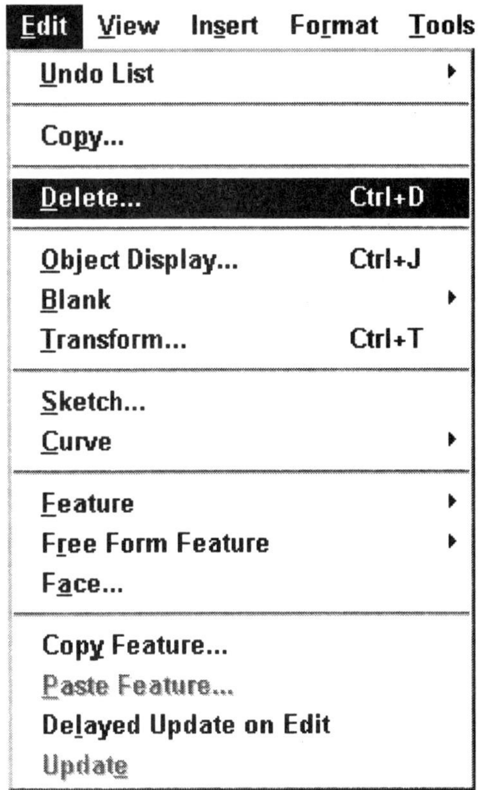

Delete:

Edit | Delete

Permanently remove objects from the Graphics window display in the Unigraphics part file.

Exercise: Delete objects

1. Select an application other than Gateway.
2. Select **Edit | Delete**.
3. Use the **Class Selection** dialog to select the objects you want to delete.
4. After the objects are selected, select **OK**, the objects are removed permanently.

Note: Restore objects by selecting **Undo List** before saving the file.

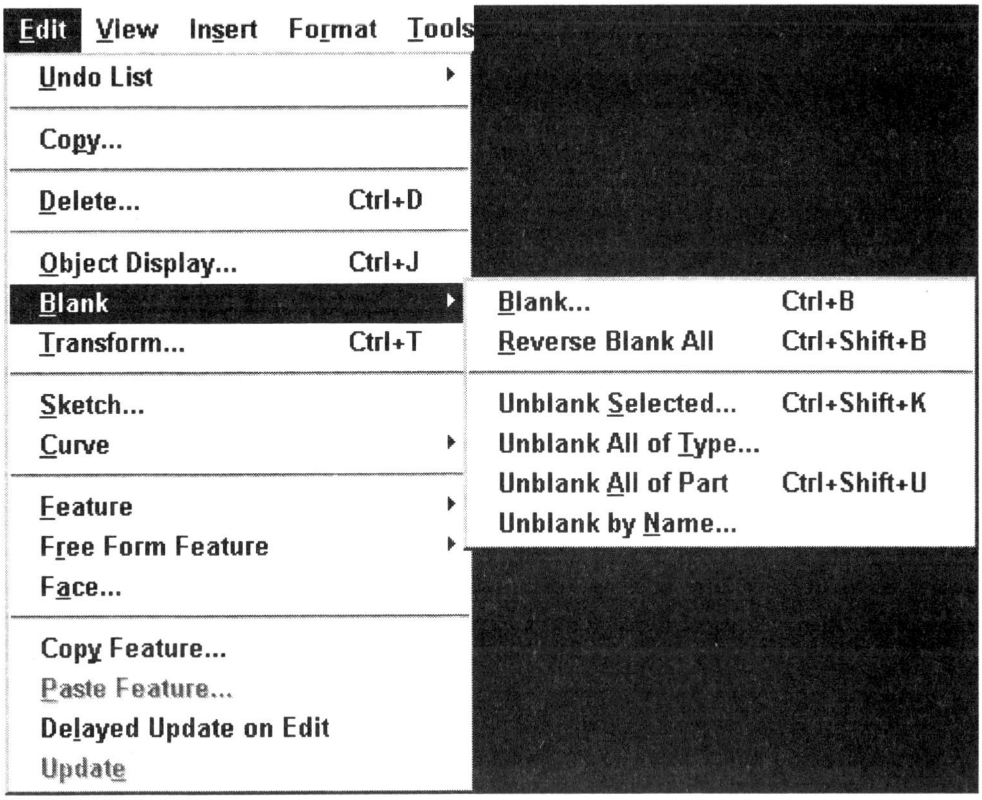

Blank:

Edit | Blank

This temporarily removes objects from view in the Graphics window display. When this is selected, the **Class Selection** dialog displays to allow the selection of objects to be blanked (removed).

Temporarily hide objects from the Graphics window display (without deleting them from the part file).
Redisplays objects hidden from the display.

Unblank Selected:

This allows the selection of blanked objects to be unblanked. When this is selected, the blank status of all objects is reversed (blanked objects are unblanked and unblanked objects are blanked). Only the previously blanked objects are displayed in the Graphics window. The **Class Selection** dialog displays to allow the selection of objects that are to be unblanked. After **OK** is selected, the blank status of all objects is again reversed. The selected objects are now unblanked.

Exercise: Blank an object

1. Select **Edit | Blank**.
2. Use the Class Selection dialog to select the objects to blank.
3. After the objects are selected, select **OK**.

Exercise: Unblank an object

1. Select **Edit | Unblank Selected**. The display changes to show all items currently blanked.
2. Select the items you want to unblank.
3. Select **OK**.

Editing Object Display:

Edit | Object Display

Color:

This allows you to change the color of the selected objects.

Exercise: Editing the color of an object

1. Select **Edit | Object Display** to display the **Class Selection** dialog.
2. Pick the objects to be edited.
3. Select **OK** to display the **Edit Object Display** dialog.
4. Select a color from the **Color** list.
5. Select **OK** or **Apply** to change the color of all the selected objects.

Layer Control:

In Unigraphics layers help you organize and manage data within a file. Every object in the database, such as a line, solid body, drafting note, or point is located on one of 256 individual layers. Layers act as a filter to specify which objects are selectable, visible, or invisible. They can be set globally to affect every view in the file, or locally to be able to show different geometry in different views. Layer settings are accessible using **Format | Layer Settings**.

Layer Settings:

The Layer Setting menu options allow you to manage the layers you want to display. The following graphics illustrate the way using layers within a design can help keep the file clean, easy to manage, and much easier to understand.

Layer Status:

Each layer can be set to one of four settings:

> **Selectable** refers to the ability to see and select objects that are on a particular layer.
> **Visible Only** allows you to display a layer in the Graphics window but you cannot select objects on that layer.
> **Invisible** turns a layer off so that it is no longer visible or selectable in the Graphics window.
> **Make Work** specifies the current **Work Layer**. Only one layer can be the work layer at any particular time. Any new geometry created in made on the work layer.

Format | Layer Settings

Define the **Work Layer** and the layers you want **Selectable**, **Invisible** or **Visible Only**. Create and edit categories and verify layer information.

Work:

This text box, located in the upper right corner, allows you to input the layer you want as the work layer. Press **ENTER** when you are finished. The specified layer of the 256 layers is assigned to be the work Layer. It is automatically visible and selectable and all newly created objects are created on the work layer. You have the option to change the work layer to any of the 256 layers but only one layer can be designated at any time.

Layer/Status/Count/Category:

This list in the dialog displays the status of layers, including which layer is the work layer and if a layer is selectable, visible only, or invisible. It can also show object count and the category name.

This lists controls the layers you want displayed in the **Layer/Status/Count/Category** list. The choices are as follows:

> **All Layers**
> All 256 layers are listed in the **Layer/Status/Count/Category** list.

Layers with Objects
Only the layers that contain objects are listed.

All Selectable Layers
Only layers that are selectable are listed.

Show Object Count:

This shows the count of objects on each layer in the **Layer/Status/Count/Category** list.

Exercise: Making a layer the work layer

1. Select **Format | Layer Settings**.
2. Type the layer number in the Work text box.
   Or
   Pick the layer you want to be the work layer and select Make Work.
3. Click **OK**.

Exercise: Changing a layer setting

1. Select **Format | Layer Settings**.
2. Pick the layer(s) you want to be selectable, invisible or visible only.
3. Select the appropriate status option, such as Selectable, Invisible or Visible Only.
4. Select **OK**.

Note: You can also change the Work Layer using the **Work Layer** toolbar.

Layer Numbers	Description
Layers 1-20	Solid geometry
Layers 21-40	Sketch geometry
Layers 41-60	Curve geometry
Layers 61-80	Reference geometry
Layers 81-100	Sheet bodies
Layers 101-120	Drafting objects
Layers 121-150	(open or company specific)
Layers 151-180	Manufacturing
Layers 181-256	(open or company specific)

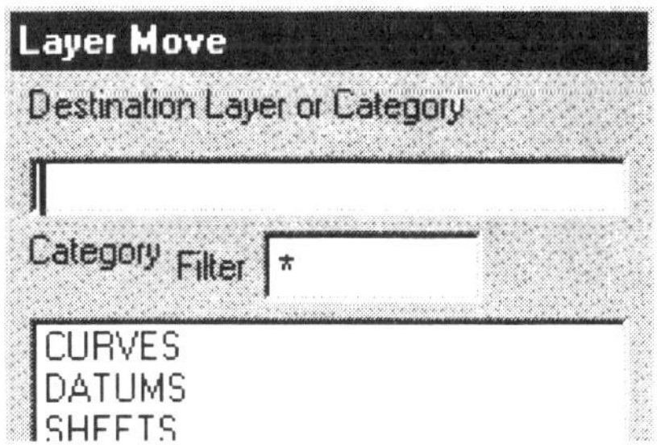

Moving or Copying Objects to Layers:

Format | Move to Layer

Move objects from one layer to another layer.

Exercise: Moving an object to another layer

1. Select **Format** | **Move to Layer** to display the **Class Selection** dialog.
2. Select the objects to be moved.
3. Click **OK**.
4. Select the destination layer or category.
5. Click **OK**.

WCS: Coordinate System Overview:

Coordinate systems define orientation. They have two properties: an origin and a rotation. The coordinate axes comprising a coordinate system are at right angles to one another. The first figure shows the coordinate axes in four basic views: Top, Front, Right and Isometric. In the top view, the X-axis is horizontal, the Y-axis is vertical, and the Z-axis is pointing toward you.

Unigraphics uses coordinate systems to define the location of objects in the database. For example, a line is defined by two endpoints. These endpoint locations are described through the use of a coordinate system.

Each part file has an Absolute Coordinate System fixed in space that defines the absolute origin of the part file. This cannot be relocated or reoriented.

Because there can be multiple coordinate systems in a part file, you must specify the system you need to work with to interpret your coordinate input data. This coordinate system is the "Work Coordinate System" or WCS. The WCS axes names differ from those of the absolute coordinate axes. The WCS axes are XC, YC, and ZC with the "C" standing for current. You can have only one coordinate system as the WCS at any given time.

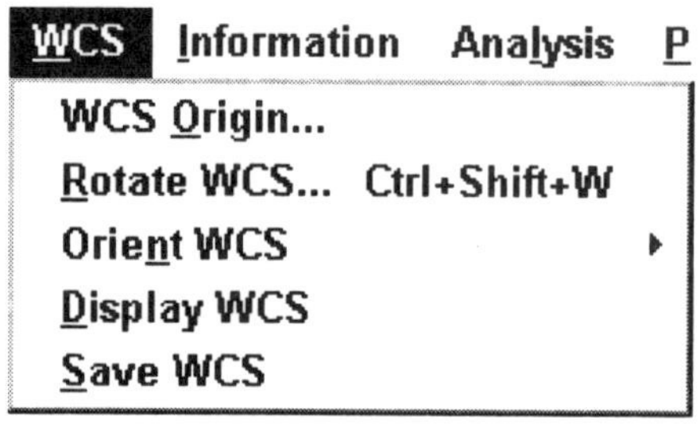

Work Coordinate System:

The Work Coordinate System (WCS) defines the orientation of new objects added to the model. It has two properties: an origin and a rotation. The **WCS** menu lets you manipulate these properties as well as saving a WCS location and making the WCS display active or inactive.

Only one coordinate system can be the WCS at any time. However, you can create, save, and retrieve as many different coordinate systems as you need in the design. These coordinate systems can have different orientations and locations. The orientation of the three axes is the "rotation" and the location of the vertex is the "origin." The XC-YC Plane of the Work Coordinate System defines the work plane on which curves are created.

WCS Origin:

This defines a new origin for the WCS. Selecting this displays the **Point Constructor** dialog to allow the definition of a new origin. After the origin point is specified, the WCS moves to the new location. The XC, YC, and ZC axes' orientation remains the same.

Rotate WCS:

This rotates the WCS about one of its axes. When this is active, the **Rotate WCS about** dialog displays. Select the axis, the rotation direction, and enter the rotation angle. Click **OK**, the WCS rotates.

Orient WCS:

This changes the orientation of the WCS. The orientation is the direction of the axes and the location of the origin.

X-Axis, Y-Axis
When you select this option, Unigraphics prompts you to pick two lines or linear edges. The coordinate system origin is at the intersection ofthe two coplanar lines/edges. The X-axis orients along the first line/edge selected with the positive direction determined by the selected end of the line/edge. The Y-axis orients along the second line/edge with the positive direction determined by the end selected. It is always perpendicular to the X-axis even if the Z-lines/edges are not.

Absolutc CSYS:

This makes the work coordinate system congruent to the absolute coordinate system.

Display WCS:

This activates and deactivates the display of theWCS in the Graphics window.

Save WCS:

This saves the WCS in the current orientation, creating a permanent saved coordinate system object (CSYS).

This coordinate system remains fixed at the saved location, even when the WCS is moved to a new origin or orientation (but not saved). To move the WCS back to the saved WCS, select **WCS | Orient WCS | Existing CSYS** and pick the saved Coordinate System.

Note: It is very important to remember which coordinate system you are working with at all times.

The **Current View** is the quickest way to orient the WCS flat to the screen in the work view.

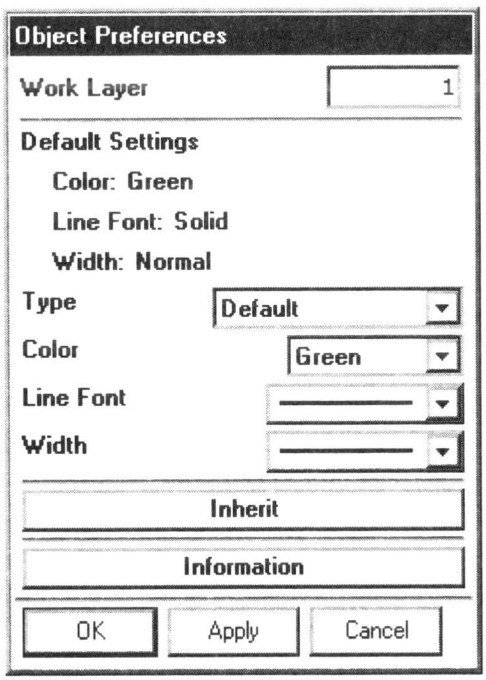

Object Display Preferences:

Use this command to set the layer, color, line font, and width for new objects. These preference settings do not affect existing objects.

Selecting **Preferences | Object** displays the **Object Preferences** dialog.

Display Mode:

This controls the way the part is displayed in each view.

- **Wireframe** displays the model as 3D curves, with every edge in the part displayed. It is sometimes faster than shaded mode (though this depends on your graphics card).

- **Fully Shaded** displays the model as fully shaded faces.

Hidden Edges:

When the Display mode is set to **Wireframe** or **Partially Shaded**, **Hidden Edges** becomes available. You can set the hidden edges in the model to be**Visible** or **Invisible**, or to display as **Thin Gray** or **Dashed** lines.

When the display of **Hidden Edges** is enabled, the view must be updated each time the part is moved or rotated. Selecting **Hidden Surface** eliminates this, updating the display whenever the part moves. However, this can be slow for large models, and the line display is of lower quality.

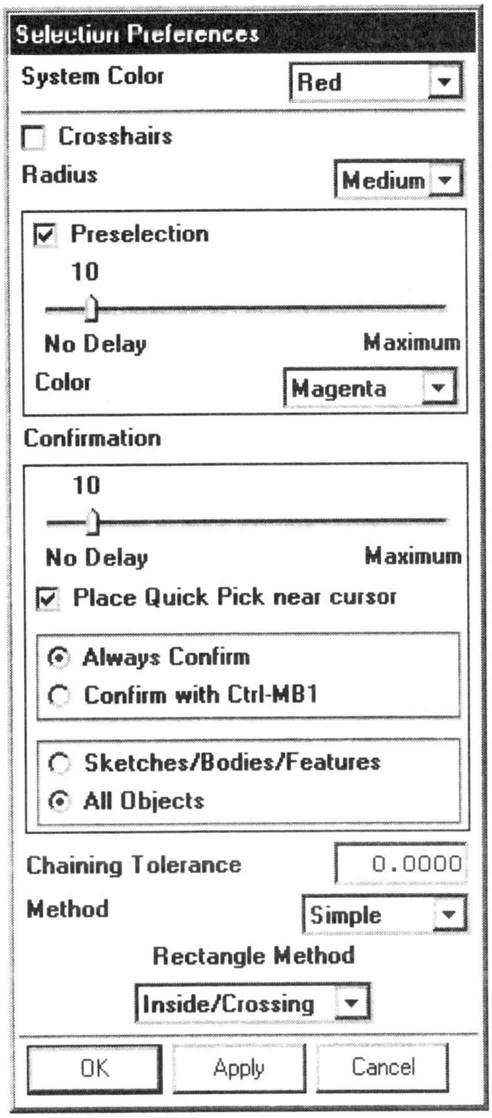

Selection Preferences:

Use this to set various options controlling the selection of objects in the Graphics window.

Select **Preferences** | **Selection** to display the Selection Preferences dialog.

System Color:

Use this to set various options controlling the selection of objects in the Graphics window.

Select **Preferences** | **Selection** to display the Selection Preferences dialog.

Crosshairs:

Controls the display of the selection crosshairs.

Radius:

Controls the radius of the selection ball (**Small**, **Medium**, or **Large**).

Summary:

In this chapter, you learned how to:

- Create a New Part File.
- Open an Existing Part File.
- Save a Part File.
- Change the Displayed Part.
- Close a Part File.
- Reopen a Part File.
- Exit Unigraphics.
- Control the Way Objects Display in Unigraphics.
- Change the Way Object Selection Behaves.
- Use the View Pop-up Menu.
- Rotate a View.
- Orient a View.
- Select Objects.
- Blank and Delete Objects.
- Edit the Object Display.
- Create Layer Settings.
- Create Visible in View Settings.
- Move Geometry to New Layers.
- Use the Absolute Coordinate System.
- Use the Work Coordinate System (WCS).
- Move the Origin of the WCS.
- Rotate and Orient the WCS.
- Save the WCS as a Coordinate System Object.

Chapter 3

Sketcher

After completing this chapter, you will be able to:

- Create a New Sketch.
- Attach a Sketch to a New Plane or Face.
- Create Sketch Curves.
- Change the Overall Preferences of the Sketcher.
- Create and Edit Geometric and Dimensional Constraints.
- Show and Remove Existing Geometric Constraints.
- Mirror Sketch Geometry.
- Convert Sketch Geometry From Active to Reference and Vice Versa.
- Position Sketches Relative to Model Geometry.
- Change the Shape of a Sketch Using the Drag Option.
- Animate the Motion of a Sketch by Changing a Sketch Dimension Value Incrementally.
- Position Features to Centerlines and Curves.
- Edit the Defining String of a Swept Feature.

Sketcher Overview:

The sketcher tool lets you create fully parametric curves called*sketch curves*. Normally 2D planar geometry, sketch curves let you create extrusion strings, sweep guides and law curves.

Sketches:

Sketches are sets of 2D fully parametric curves. First creating base curves and then adding geometric and/or dimensional constraints to control the shape of the curves create sketches. By adding these constraints to the sketch curves you can capture design intent and later modify the sketch shape.

Normally you use sketches to define sweep section strings, guide curves or law curves. They are excellent tools for developing non-standard shapes where standard features do not work.

An advantage in using the sketcher is to create sketchcurves on a datum plane or solid planar face. If a datum plane or face changes orientation or moves, the sketch geometry stays attached to the plane or face. You can also add dimensional and geometric constraints to the sketch. Using the dimensional constraints, you can create system-defined expressions and then link them to each other in the sketch or to other features of the solid.

You can use the sketcher in many different applications requiring non-standard shapes. There are occasions where it may become difficult to create the shape you desire by using combinations of form features. In these cases, you can create a sketch to define the geometry. The graphics illustrate some applications for sketching.

Sketch Placement:

You can attach a sketch to a planar face or a datum plane. The sketch stays associative to the attachment face. If the face changes orientation or location, the sketch changes with it.
When a new sketch is created that is not attached to a solid face, Unigraphics creates a fixed datum plane and two fixed datum axes, which represent the horizontal and vertical directions of the sketch.

Sketch Management:

Organizing the data within a CAD file is always important when many different individuals are potentially using the data. Data organization is especially important when using sketches because you can define sketch information on many layers and views, with different sketch names. Depending on the complexity of a design, a single solid model can easily contain over 20 sketches. With this amount of data, it is prudent to use good techniques to manage your sketches.

Note: Put each sketch (and related datum axes and planes) on a separate layer to make it easier to find information.

Always set the work layer before creating a sketch. Although you can move objects, it is easier to create objects on the correct layer.

Give the sketches meaningful names. For example, if you create a sketch for the outer profile of a part, name the sketch **s_{layer number}_profile**.

Create a set of base datum planes early in the model creation process. Use these datum planes to orient and position features and sketches. You must order the datums before the sketches in the feature order list.

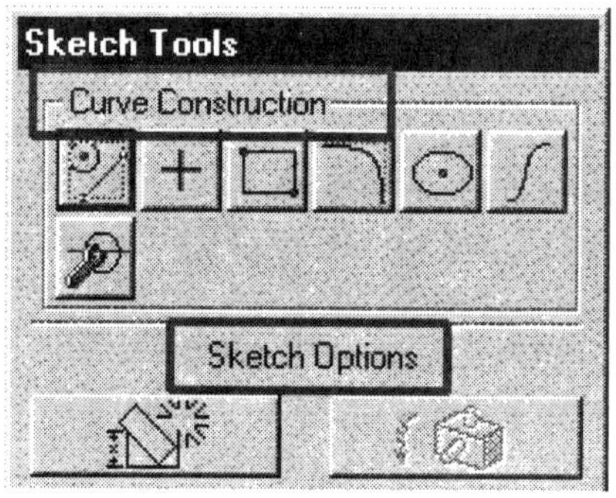

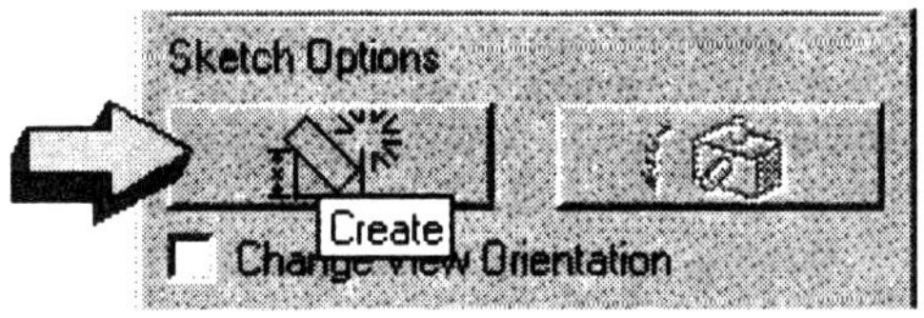

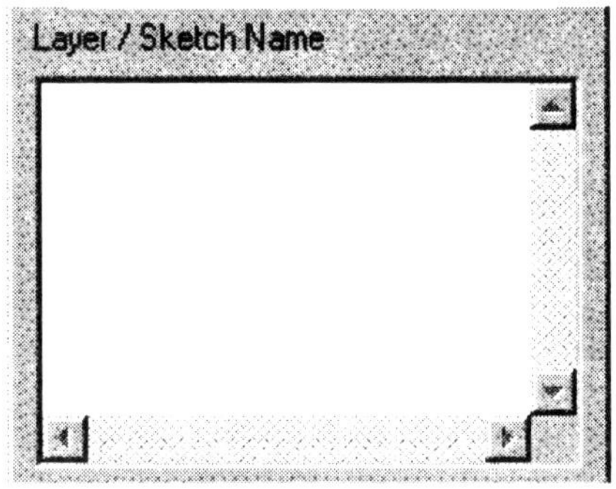

Creating a Sketch:

 **Insert | Sketch**

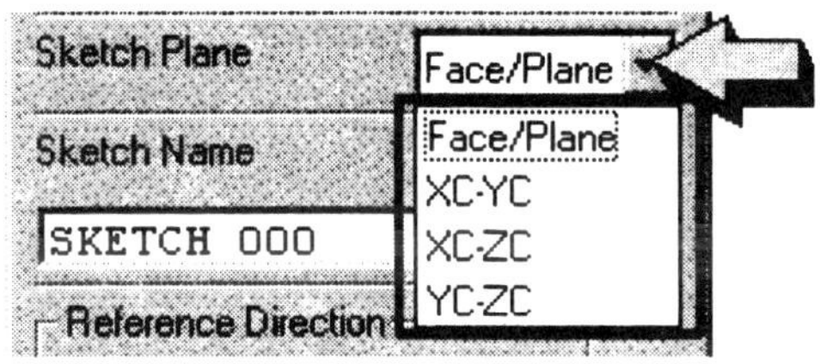

Change View Orientation:

This controls the sketch view orientation. When this is checked and a sketch is created or activated, the orientation of the view squares up to you based upon your selections for the Sketch Plane and the Reference Direction. When this is not checked, the view does not change to match the sketch.

Create:

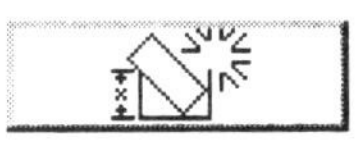 Click Create, the Create Sketch dialog displays.

The Selection Steps icons are only selectable if a solid or datum plane exists in the part file and Sketch Plane is set to Face/Plane.

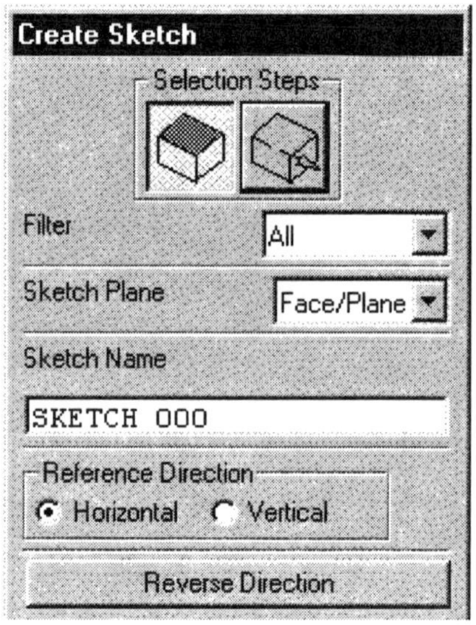

Sketch Plane:

If Sketch Plane is set to Face/Plane the Attachment Face/Plane step will allow you to select an existing datum plane or planar face from the model to be theattachment face for the sketch.

Sketch Plane
This filter determines the type of planar entity the sketch is attached to.

You can select any planar solid face, datum plane, or principal plane of the WCS. After you pick a placement face, a direction arrowappears to indicate the orientation of the plane. When the view is oriented to the sketch plane, this arrow points out of the screen. If a datum plane is used as the sketch placement face, click Reverse Direction, if needed, to reverse the direction arrow.

Reference Direction

When attaching a sketch to a datum plane or planar face, you can specify a Horizontal or Vertical direction to constrain the orientation of the sketch. Select datum planes, planar faces, or linear edges to define the direction. A direction arrow appears to indicate the positive direction of the sketch. When the view is oriented to the sketch plane, the Horizontal Reference Direction arrow points in the +XC direction, and the Vertical Reference Direction arrow points in the +YC direction. Selecting Reverse Direction can reverse the direction arrow (and the orientation of the sketch).

Sketch Name
The default sketch names are sequential beginning with "SKETCH_000" and you can overwrite them with meaningful names (i.e., s_{layer number}_name). The prefix "SKETCH_" is controlled by a setting sketch preference.

Exercise: Creating an unattached sketch

1. If not already, start **Unigraphics**.
2. Create a file named **sketch_1**.
3. Enter the modeling application.
4. Select **Insert | Sketch**.
5. If desired, activate **Change View Orientation**.
6. Click **Create**.
7. Select **XC-YC**, **XC-ZC**, or **YC-ZC** from the **Sketch Plane** list.
8. Enter a **Sketch Name**.
9. Click **OK** or **Apply** to create the sketch.

Exercise: Creating an attached sketch

1. Open the file, **sketch_2**.
2. Select **Insert | Sketch**.
3. If desired, activate **Change View Orientation**.
4. Click **Create**.
5. Click **Attachment Face/Plane**.
6. Pick a solid planar face or datum plane.
7. Click **Reference Direction**.
8. Set **Reference Direction** to **Horizontal** or **Vertical**.
9. Pick a planar face, datum plane, or linear edge.
10. Select **Reverse Direction** if necessary.
11. Enter a **Sketch Name**.
12. Click **OK** or **Apply** to create the sketch.

Note: You can create an unlimited number of sketches per part file.

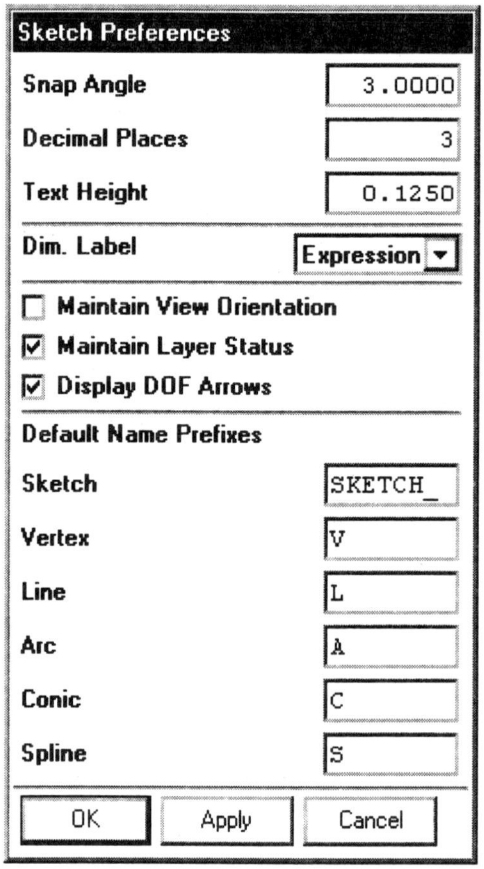

Sketch Preferences:

There are several preferences you can set to help control the Sketcher functions and how sketch curves display in the Graphics window. These preferences are set by selecting **Preferences | Sketch**.

Snap Angle:

This value lets you control whether a line created is horizontal, vertical oras drawn. This value is the angle a line can vary from either side of a horizontal or vertical orientation in a sketch and still be considered horizontal or vertical.

Indicate the start point for a line and then specify the second point for the line. Ifthe line falls within this specified range, it automatically adjusts to a horizontal or vertical line. This applies if the second point is a cursor location point. Any line that uses endpoints of an existing curve as the second point, and is created within this range, does not automatically become horizontal or vertical. To create a line at an angle less than the snap angle, create it at an exaggerated slope and add a dimension to the line to constrain the angle of the line.

Decimal Places:

This controls the number of decimal places a sketch dimension displays. The number of displayed decimal places does not affect the accuracy.

Text Height:

This determines the character size of the sketch dimensions.

Display DOF Arrows:

Check this to display the degree-of-freedom (DOF) arrows. It is a good idea to always display the degree-of-freedom arrows. This indicates unconstrained portions of the sketch.

Sketch Curve Construction:

Use sketch curve geometry for extrusions, guide strings, scaling strings, and law curves. If a sketch is activated, you can create curves using the Sketch Tools Curve Construction options. These curves are drawn on the sketch plane, and are controlled using dimensional and geometric constraints to determine their size and relationship to adjacent curves. The sketch Curve Construction options are located at the top of the Sketch Tools dialog. All the curve creation options in this section are similar to those available when you select Insert | Curve.

Basic Curves:

 This is used to create lines, arcs, circles and fillets and to trim and edit the parameters of a curve.

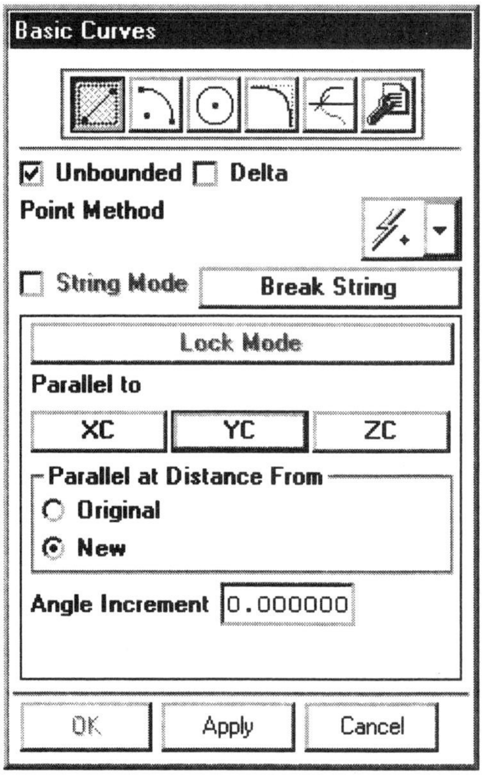

Point:

A point is the simplest element in modeling. It is defined by a set of X, Y, and Z spatial coordinates.

This creates a point using the Point Subfunction dialog.

Line:

A line is the least complex curve type and is defined by a start point and an end point.

Circle/Arc:

Unigraphics defines a circle by a center point and a diameter. The start and end point of a circle is coincidental. An arc is defined either by a start point, end point and a point on the arc, *or* a center point, start point, and end point. The term "arc" is often refers to a portion of a circle.

Rectangle:

Creates a rectangle by specifying two diagonal points.

Fillet:

Creates a simple fillet or a 2-curve fillet. The 3-curve fillet found in the Basic Curves dialog as well as the Trim Options are not available while a sketch is active. Fillet automatically makes the fillet tangent and trims the adjoining curves.

Simple Fillet:
This creates an arc between two co-planar non-parallel lines and automatically trims the geometry.

2-Curve Fillet:
This creates an arc between two curves and automatically trims the geometry. You must pick the curves in counterclockwise order with respect to the WCS.

Radius:
Input the size of the fillet to create.

Ignore Radius:
This is only available with the 2-Curve fillet option. When activated, this ignores the value in the Radius text box and uses the pick location of the first curve as the start point of the fillet. The second curve selected determines the fillet size by creating the fillet tangent to both curves and through the start point.

Inherit:
This allows you to select an existing arc to extract the radius value from.

Ellipse:

Create an ellipse by specifying the center of the ellipse, the major and minor axis, the start and end angles and the rotation angle of the ellipse.

Polygons:

Insert | Curve | Polygon

Number of Sides:

This determines the number of sides in the polygon. If the number of sides exceeds 513, an error message displays.
Polygons must have three or more sides.

Inscribed Radius:

This determines the size of the polygon by specifying the radius of the largest circle that fits inside the polygon.

Side of Polygon:

This determines the size of the polygon by specifying the length of one of the sides. The length of all sides is equal.

Circumscribed Radius:

This determines the size of the polygon by specifying the radius of the circle passing through all of the polygon's corners.

Orientation Angle:

This defines how the polygon is orientated on the XC-YC plane. The orientation angle is measured counterclockwise from the positive XC-direction.

Note: Use **Polygon** with an orientation angle instead of **Rectangle** if you need a rectangle that is not parallel to the XC-YC axes.

Creating Chamfers:

 **Insert | Curve | Chamfer**

Simple Chamfer:

This applies a 45-degree chamfer at a specified offset value when the chamfer is applied between two perpendicular lines. If you create a chamfer between two lines that are not perpendicular, the chamfer uses the offset value along both curves.

When selecting the corner, both lines must pass through the selection ball with the crosshairs on the side of the chamfer.

User Defined Chamfer:

This creates a chamfer between any two coplanar curves.

Automatic Trim trims both curves to the chamfer curve.

Manual Trim asks whether to trim each curve. If you select **Yes,** you must select the end of the curve to trim away. If you select **No** for either curve, it remains unchanged.

No Trim leaves both curves untrimmed.

Offset and Angle:

If you know the offset and the angle, enter the values in the **Offset** and **Angle** text boxes. The offset is measured perpendicular to the first curve selected and the angle is measured from the second curve selected.

Offset Values:

If you know both offset values, click **Offset Values** and enter the two offsets. The first offset is measured perpendicular to the first curve selected and the second is measured perpendicular to the second curve.

Constraints:

Constraints are powerful tools that allow you to control the way sketch geometry varies. You can control the design intent of your sketches by the application and manipulation of these constraints.

Constraints are divided into two types, **Dimensional** and **Geometric**.

Dimensional constraints assign a numerical value to the sketch geometry, and may define things like the length of a line, the radius of an arc, or the distance between a two items.

Geometric constraints assign a physical relationship between sketch entities, such as whether a line and an arc are tangent to each other, whether two lines are perpendicular to each other, or whether three arcs all have the same radius.

Constraints can be applied to geometry in the sketch as well as relative to other existing geometry in the model, such as solid edges and datum features. Dimensional constraints can be used to position the sketch relative to a corner of a solid model, and Geometric constraints can be used to limit a sketch curve to always intersect with a solid edge.

Constraints Dialog and Status Line:

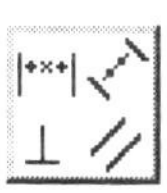
Once a sketch is created, dimensional and geometric constraints can be added. When you click Constraints, found in the Sketch Tools dialog, your sketch appears with Degree of Freedom (DOF) arrows that mark the geometry needing to be constrained. The

Status line displays the number of geometric and dimensional constraints remaining for an unconstrained sketch. At this point, the Constraints dialog is also visible.

As you assign constraints, the DOF arrows disappear and the number of constraints required decreases until the status line indicates the sketch is fully constrained.

Note: You are not required to fully constrain a sketch, but it is a very good idea.

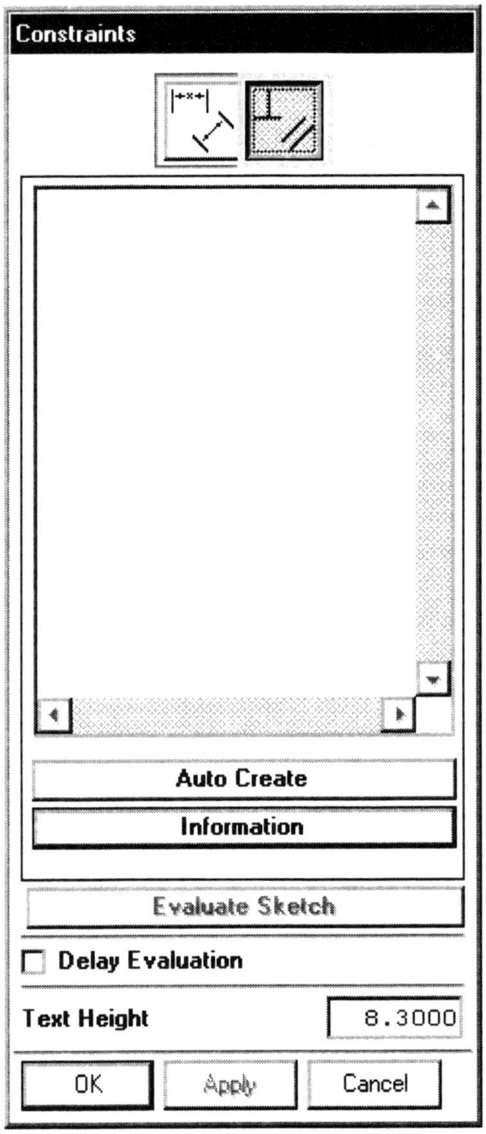

Geometric Constraints:

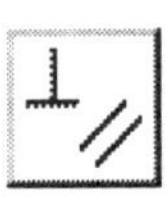 Assign physical relationships between sketch geometry and other part entities.

Geometric Constraint Types:

The following geometric constraint types are available:

Fix fixes a point in space in the exact location where it is created. Apply this constraint to features such as points, endpoints of curves, and arc centers.

Tangent creates a tangent condition between two sketch curves. Apply this constraint between an arc and a line or two arcs.

Horizontal forces a line to be parallel to the horizontal reference direction of the sketch. Apply this constraint to lines.

Vertical forces a line to be parallel to the vertical reference direction of the sketch. Apply this constraint to lines.

Parallel assigns a parallel condition between two or more lines. Apply this constraint to lines.

Perpendicular assigns a perpendicular condition between two or more lines. Apply this constraint to lines.

Collinear forces two or more lines to be collinear. Apply this constraint to lines.

Equal Length forces two or more lines to be equal length. Apply this constraint to lines.

Constant Length forces a line to remain at a fixed length. Apply this constraint to lines.

Constant Angle
Forces a line to remain at a fixed angle from the horizontal reference direction of the sketch. Apply this constraint to lines.

Point on Curve forces a point to lie on a curve. Apply this constraint to features such as points, endpoints of curves, or arc centers. Select a line, arc, solid edge, or datum feature to use as the constraining curve.

Midpoint of Curve forces a point to lie on the midpoint of a curve. Apply this constraint to features such as points, endpoints of curves, or arc centers. Select a line, arc, or solid edge to use as the constraining curve.

Coincident forces two points to share the same location. Apply this constraint to features such as points, endpoints of curves, or arc centers.

Concentric forces two or more arc centers to share the same location.

Equal Radius forces two or more arcs to have the same radius.

Mirror forces two objects to maintain a mirrored image of each other. Apply this to curve entities.

Point on String lets you pick the endpoint of a sketch curve and constrain to an extracted curve.

Scale, Uniform scales a spline proportionately between its knot points when the overall length or height is changed.

Scale, Non-Uniform keeps all knot points at their current location when changing the overall length or height of the spline.

Slope constrains the selected spline knot point to the slope of a selected reference curve.

Delay Evaluation:

When this active, the sketch updates later.

This activates Evaluate Sketch. To force the sketch to update, click Evaluate Sketch.

Exercise: Adding geometric constraints to sketches

1. Open the file, **sketch_3**.
2. Enter the Modeling application.
3. Select **Insert | Sketch**.
4. Activate the sketch you want to work in by selecting it from the list and click **Activate**.
5. 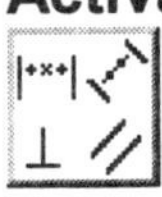From the **Sketch Tools** dialog click **Constraints**.
6. This displays the **Constraints** dialog.
7. 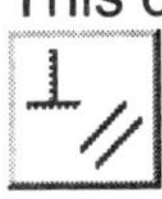Click **Geometric** to update the dialog.
8. Pick the geometry that you need to constrain in the Graphics window. The constraints are available to use are based on the type of geometry selected.
9. Select the applicable constraint types from the list.
10. Select **OK** or **Apply** to update the geometry to match the new constraint.

Note: After selecting a constraint, the Apply button is in a green box, signifying the ENTER key can be used to Apply the constraint.

Make Delay Evaluation active when modifying the dimensions of a sketch to prevent the sketch from changing shape.

More than one constraint type can be applied in a single step. For example, two lines can have a Parallel and an Equal Length constraint applied in the same step. Press CTRL to make multiple selections from the constraints list.

Be careful in the way you make entity selections. If you need to create a constraint that does not constrain a point, be sure to pick the entity away from the end or control points.

Show Remove Constraints:

This displays geometric constraints and provides a way to remove them.

When you click Show/Remove Constraints, the Show/Remove Constraints dialog displays.

If you are in this dialog and move your cursor over sketch curves in the Graphics window, geometric constraints display graphically as constraint markers whenever the cursor touches a sketch curve that has geometric constraints applied toit. You can also display the constraints in the list by picking a sketch curve from the Graphics window.

All in the Current Sketch:

Click this to list all of the geometric constraints in the current sketch.

Remove Highlighted Constraints(s):

Use this to remove a constraint. Select to highlight a constraint(s) in the list and click Remove Highlighted Constraint(s) to remove the constraint.

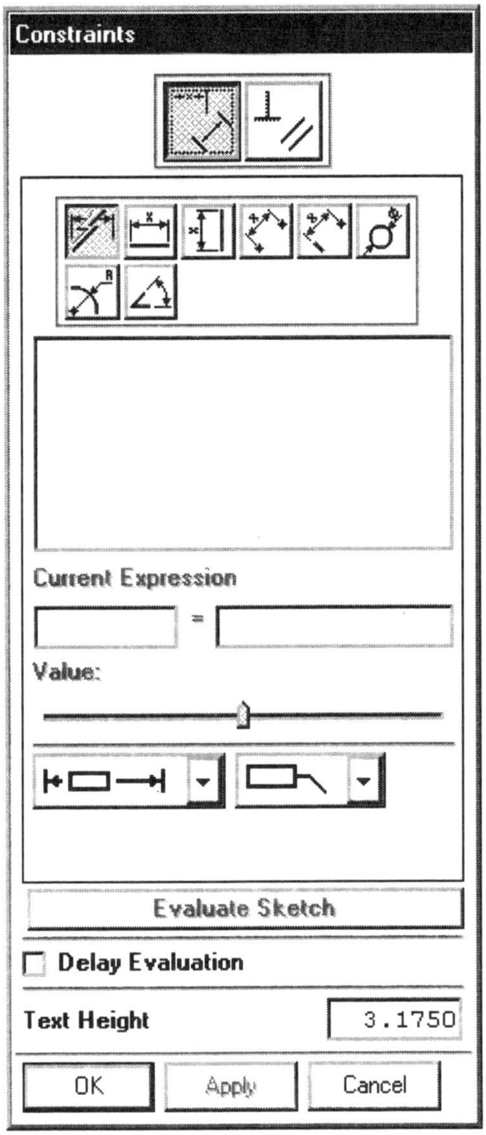

Dimensional Constraints:

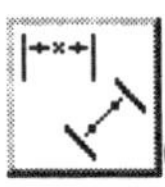Control the size of sketch curves.

Apply constraints between an edge and a sketch curve or a datum feature and a sketch curve.

Dimensional Constraint types:

Inferred:

 This creates dimensional constraints based on inferred locations or smart picks.

Horizontal:

 Constrains two points parallel to the X-axis of the sketch. Only pick the line once to dimension a line as horizontal.

Vertical:

 Constrains two points parallel to the Y-axis of the sketch. Only pick the line once to dimension a line as vertical.

Parallel:

 Constrains two points radially (the shortest distance between two points). Only pick the line once to dimension a line as parallel.

Perpendicular:

 Constrains a point to a line at a normal (shortest distance between the line and the point). When using this option, always pick the line first and then the point.

Diameter:

 Constrains the diameter size of a circle.

Radius:

 Constrains the radius size of an arc.

Angular:

 Constrains an angle between two lines.

Expression List:

Dimensional constraints create expressions that can be edited while in theConstraints dialog or the Edit Expression dialog. After placing a sketch dimension, the current dimensional value displays in the Constraints dialog as an expression.

If the value changes, the sketch immediately changes shape. To edit a dimension's value, select it from the list and type in a new value or use the slider bar to change the value dynamically.

Dimension Placement:

Dimensions can be placed using Auto Placement, Manual Placement, Arrows In and Manual Placement, Arrows Out.

Leader Placement:

The leaders can be located using Leader from Left and Leader from Right.

Text Height:

You can control the size of the text by entering values in the Text Height text box (located at the bottom of the Constraints dialog).

Exercise: Adding dimensional constraints to sketches

1. Continue using **sketch_3.**
2. Application **Modeling.**
3. Activate the sketch.
4. 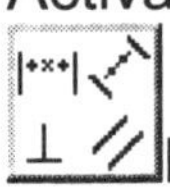In the **Sketch Tools** dialog click **Constraints**.
5. This displays the **Constraints** dialog.
6. 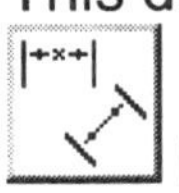 Ensure the **Dimensional** icon is selected.
7. Pick the dimension type you need to place.
8. Pick the curves to dimension from the Graphics window.
9. The dimension preview drags with the cursor. Click to place the dimension.
10. Select **OK**, **Apply** or **Cancel** to close the **Constraints** dialog.

Note: To change the origin of the sketch dimension, pick the dimension in the Graphics window twice. On the second pick, hold down MB1 and drag the cursor to a new location. The dimension follows.

You can edit the value of a sketch dimension by selecting it from the list and dragging the slider bar in the middle of the dialog.

Constraint Colors:

The color of a constraint in the Graphics window indicates certain information about the constraint. The following table explains the meaning of each color.

Note: If you are in a situation in which adding a dimension or editing a dimension has caused either a yellow or pink constraint color, undo is available to let you to go back to the previous state of the sketch.

Mirror:

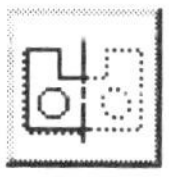 Creating a sketch that is symmetrical about a centerline.

Mirror Centerline:

 Use this to pick a linear curve that acts as the centerline for the mirroring operation. The line selected is converted to a reference line when the operation is complete.

Mirror Geometry:

 Use this to pick the geometry to mirror about the centerline. Do not include the centerline when selecting the objects to mirror.

Exercise: Mirroring sketches

1. Open **sketch_4**.
2. Enter the Modeling application.
3. Select **Insert | Sketch**.
4. Activate the sketch.
5. 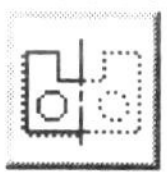 Click **Mirror**
6. The **Mirror Sketch** dialog displays.
7. Pick a linear curve from the sketch to mirror about.
8. Click **OK** or 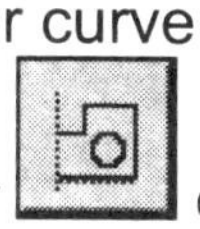 click **Mirror Geometry** to move to the geometry selection step.
9. Pick the geometry to be mirrored.
10. Click **OK** or **Apply** to complete the mirror operation.

Note: Can you mirror more than one curve at a time?
Yes, you can select multiple curves but you can only select one mirror centerline.

Notice that when the Mirror Centerline is an active curve, Unigraphics automatically converts the curve to a reference line when the mirror is completed. If you want this curve to be active you must convert it back after the mirror is completed.

Convert to/from Reference:

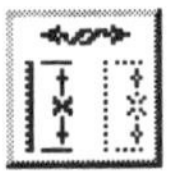 Switch sketch curves and dimensions between Reference and Active status.

The default for the sketcher is to create all curves and dimension as active. This gives you the ability to convert them to Reference (and back to Active if necessary). Reference curves are gray phantom curves and are not included when the sketch is used to create solid geometry. Reference dimensions are white and are not utilized to constrain the sketch. When the Convert To/From Reference icon is selected the following dialog displays.

Filter:

Use this list box to select curves or dimensions when picking objects to convert.

Exercise: Converting sketch curves to/from reference curves

1. Continue working with **sketch_4**.
2. Select **Insert | Sketch**.
3. Activate the sketch.
4. 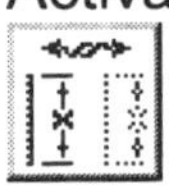 Click **Convert To/From Reference**.
5. Click **Reference** or **Active**.
6. Pick the geometry to convert in the Graphics window.
7. Click **OK** or **Apply** to complete the process.

Note: You can select more than one curve or dimension at a time to convert.

Relative Position:

A Relative Position is a relationship between a sketch and an existing solid, curve type or datum plane.
You can position any sketch attached to a solid face or datum plane relative to a curve, solid edge or datum plane. Use the Positioning Dimensions icon to access the Positioning dialog.

The process of positioning a sketch is identical to the positioning of form features. A relationship is created between a curve, solid edge or datum plane and a sketch point or curve. Pick a curve, edge or datum plane as the target edge and then pick a sketch curve. Using the positioning dimensions is one more way to define design intent.

When the curve, solid edge or datum plane is revised, the sketch moves accordingly. You

can change these relationships by deleting and reapplying positioning dimensions or by editing the positioning dimension values.
Relative Positioning versus Applying Dimensional Constraints:
When relative positioning dimensions are used to position a sketch you are positioning the entire sketch relative to the model just as if it was a form feature. If a dimensional constraint is applied between a sketch curve and a solid edge or datum plane using the dimensional constraint, you are only positioning that particular curve relative to the edge or datum plane and not the entire sketch.

Positioning Dimensions:

 Position a sketch relative to an existing curve, solid edge or datum feature.

Positioning Types:

When positioning sketches, the sketch being positioned is referred to as the "tool object" and the solid edge or datum feature that it is being positioned to is the "target object."

Horizontal:

 This assigns a positional constraint that is aligned parallel to the horizontal reference or 90-degrees from the vertical reference.

Vertical:

 This assigns a positional constraint that is 90-degrees to the horizontal reference and aligned parallel to the vertical reference.

Parallel:

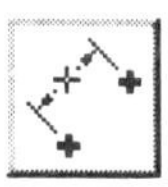 This assigns a parallel positional constraint between two points, using entity endpoints, arc center points or arc tangent points. The dimension line is always parallel to the imaginary line created between the two points selected. Use Parallel to create a radial dimension.

Perpendicular:

 This assigns a positional constraint perpendicular from a linear edge of the target solid to a point of the sketch or tool solid. Datum planes, datum axis, or an existing curve can also be used to define the target edge.

Parallel at a Distance:

 This assigns a positional constraint to be parallel from a linear edge of a target solid to a linear edge of a sketch or tool solid.

Angular:

 This assigns an angular positional constraint between a linear edge of the target solid or reference datum to a linear edge of the feature or sketch.

Point onto Point:

 This assigns a constraint dimension just like the Parallel constraint but with the fixed distance between the two points set to zero.

Point onto Line:

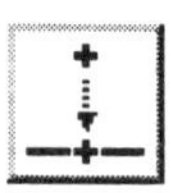 This assigns a constraint dimension the same as Perpendicular, but with the fixed perpendicular distance between the edge and point set to zero.

Line onto Line:

 This assigns a constraint dimension just like the Parallel at a Distance constraint but with the distance between the linear edge of the feature or sketch and the edge on the target solid set to zero.

Exercise: Adding positioning dimensions to a sketch

1. Open **sketch_5**.
2. Select **Insert | Sketch**.
3. **Activate** the sketch.

4.  Click **Positioning Dimensions.**
5. The **Positioning** dialog displays.
6. Select the desired dimension type to apply to the sketch, the corresponding dialog displays.
7. Pick the **Target Edge** for the dimension from the model.
8. Pick the desired sketch curve to dimension to.
9. Type in a value for the dimension.
10. Click **OK** to place the dimension and return to the **Positioning** dialog.
11. Repeat Steps 5-8 to place more dimensions.
12. Click **OK** from the **Positioning** dialog when you finish placing dimensions.

Note: If you create a dimensional constraint between an edge and a sketch curve, an error message displays if you attempt to add positioning dimensions using the process defined above.

You can modify the (0,0) dimension for the constraints Point on Point, Line onto Line and Point onto Line later so they are not 0.

To edit sketch positioning dimensions select **Edit | Positioning** and select the sketch feature from the list.

Drag:

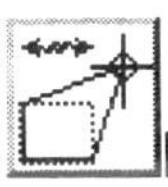Dynamically modify dimensional constraints in a sketch.

Dynamically modify under-constrained curves in a sketch.

Exercise: Modifying a dimensional constraint using drag

1. Create a sketch with closed geometry.
2. Select **Insert | Sketch**.
3. 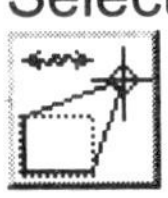 Click **Drag**.
4. The **Drag** dialog displays.
5. In the Graphics window, pick the dimensional constraint you need to modify.
6. Edit the dimension by entering an **Increment Size** and clicking the right and left arrow buttons in the **Drag** dialog. *OR*
   Click and drag the cursor in the Graphics window, releasing MB1 when the geometry is in the desired location. The geometry changes dynamically based on the sketch constraints, and the dimension updates appropriately.
7. Click **OK**, **Back** or **Cancel** to exit the **Drag** dialog.

Exercise: Modifying curves using drag

1. Select **Insert | Sketch**.
2.  Click **Drag**.
3. The **Drag** dialog displays.
4. Pick the sketch curves to drag.
5. Click **Done Selecting**.
6. Click and drag the cursor in the Graphics window. The geometry changes dynamically based on the sketch constraints. Release MB1 when the geometry is in the desired location.
7. Click **OK**, **Back** or **Cancel** to exit the **Drag** dialog.

Note: Can I move a whole sketch using the Drag method? Yes, if you select all the sketch curves (and there are no positioning dimensions to hold the sketch in place), you can move the whole sketch at one time.

If you drag a dimension by clicking and moving it in the Graphics window, you cannot place it exactly. If necessary, clean up this dimension using the Constraints dialog or Expression Editor.
Drag is a good method to use in order to get a sketch close to the size you desire. Once close, you can then go back and add dimensions and geometric constraints to make the sketch precise.

Animate:

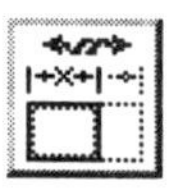 Incrementally change a Dimensional constraint within a given range.

Upper and Lower Limit:

Dimensional constraints display in a list window on the Animate dialog. When you select a constraint, Unigraphics assigns an initial Lower Limit and Upper Limit. The lower limit is 90% of the constraint value and the upper limit is 110% of the constraint value. You can adjust these limits using these text boxes. A "cycle" is defined as the motion from the nominal value to either the Upper Limit or Lower Limit.

Steps/cycle:

Steps/Cycle is the number of animation steps per cycle. This value must be greater than zero.

When you click Apply to begin the animation the sketch changes shape dynamically as the dimensional constraint changes in increments. This is a good visualization technique to illustrate the effects of the constraints.

Exercise: Using the animate option

1. Open **sketch_6**.
2. Select **Insert | Sketch**.
3. 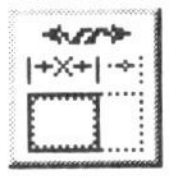 Click Animate.
4. Select a dimensional constraint from the list or from the Graphics window.
5. If needed, in the Animate dialog type the appropriate values in the Lower Limits, Upper Limits and Steps/Cycle text boxes.
6. Click Apply to start the animation. The sketch dynamically changes.
7. Click OK on the Stop Animation dialog to stop the animation.
8. Select Cancel to exit the Animation dialog.

Note: Does animating a dimension actually change the dimension value?
No. Once the animation is stopped, the sketch dimension reverts back to its original value.

The Animate function can help determine why a sketch would become over-constrained. Animate the dimension in question until the sketch becomes over-constrained. You can then evaluate the geometry to determine the failure.

Edit Defining String:

Sketches are two-dimensional profiles selected as section strings and swept, revolved or extruded into solid geometry. Edit Defining String allows you to add and remove objects to and from an existing swept feature's defining string.
Because designs are always changing, there are occasions in which you need to add curves to or remove curves from a sketch. If the sketch curves are being referenced in a swept feature, you usually need to incorporate these changes in the swept feature. The Edit Defining String allows you to do this without having to delete the swept feature and recreate it from scratch.

Editing the Defining String of a Swept Feature:

 Select the Edit Defining String icon to display the Add/Remove dialog. Geometry that is part of the swept feature section string is highlighted. To add geometry to a string, pick the geometry in the Graphics window. To remove geometry, press SHIFT and pick the geometry in the Graphics window. It no longer highlights when you remove it from the section string definition. Use the Filter list to eliminate some of the geometry choices.

When the necessary edits have been performed, click OK and then click Update Model from the Sketch Tools dialog to update the solid.

Note: Can I select objects from another sketch to add to the defining string of a swept feature? No.

After sketch curves are swept or extruded, you cannot delete them until they are removed from the swept features string definition.

Offsetting Sketch Curves:

 Offset extracted complex curves in a sketch for extrudes, sweeps, and guide strings.

Exercise: Offsetting extracted curves in a sketch

1. Open **sketch_7**.
2. Select **Insert | Sketch**.
3. **Activate** the sketch.
4.  Click **Offset Extracted Curves.**
5. Set **Offset by** to either **Distance** or **Law Control**.
6. Enter a value in the **Distance** text box if **Offset by** is set to **Distance**.
7. Click **Reverse Direction** to reverse the direction the offset is applied.
8. Click **OK** to create the offset geometry.

Note: Offset extracted curves do not need constraints applied to them.

Advanced Curve Types:

Parabola:

**Insert | Curve | Parabola**

Create a parabola through a set of points equidistant from a point (the focal point) and a line (the directrix).

A parabola is a smooth curve because it has no reversals of curvature. This type of curve is useful for creating a fillet between two curves, or as a starting shape. Because you cannot control the slope at the parabola's endpoint, start with the parabola and then create the remaining curves.

Focal Length is the distance from the vertex point to the focal point. This value must be greater than zero.

Minimum DY is the distance from the parabola's start point to the axis of symmetry. This value can be positive or negative. A negative value places the start of the parabola on the negative side of the axis of symmetry.
Maximum DY is the distance from the end point of the parabola to the axis of symmetry. This value can be negative or positive.

Rotation Angle is the angle of the axis of symmetry relative to the +XC axis.

Exercise: Creating a parabola

1. Create a part file **parabola**.
2. Select **Insert | Curve | Parabola**.
3. Define the parabola's vertex to display the **Parabola** dialog. The vertex is the intersection of the parabola and the axis of symmetry.
4. Enter the **Focal Length**, **Minimum DY**, **Maximum DY** and **Rotation Angle** parameters.
5. Click **OK** to create the parabola.

Hyperbola:

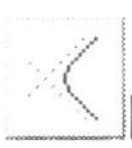**Insert | Curve | Hyperbola**

Semitransverse is the distance from the center point of the hyperbola to the apex, measured along the axis of symmetry. It defines the apex of the hyperbola.

Semiconjugate is the distance from the apex to the asymptote, measured perpendicular to the axis of symmetry. The **Semiconjugate** and **Semitransverse** define the slope of the hyperbola.

Minimum DY is the distance from the hyperbola's start point to the axis of symmetry. It is measured perpendicular to the axis of symmetry, and can be positive or negative. A negative value places the start of the hyperbola on the negative side of the axis of symmetry.

Maximum DY is the distance from the hyperbola's end point to the axis of symmetry. It is measured perpendicular to the axis of symmetry, and can be positive or negative. This value defines the end point of the hyperbola.

Rotation Angle is the angle of the axis of symmetry relative to the +XC axis.

Exercise: Creating a hyperbola

1. Create a part file **hyperbola**.

2. Select **Insert | Curve | Hyperbola**.
3. Define the center point of the hyperbola to display the **Hyperbola** dialog.
4. Enter the **Semitransverse**, **Semiconjugate**, **Minimum DY**, **Maximum DY**, and **Rotation angle** parameters.
5. Click **OK** to create the hyperbola.

Helix:

 **Insert | Curve | Helix**

Number of Turns determines the number of full 360-degree turns in the helix.

Pitch determines the pitch of the helix. If the pitch is 0, Unigraphics creates a planar helix (a 2D spiral).

Radius Method determines how the radius definition forthe helix.

> **Use Law**
> This uses a radius defined by law. Use this to create a tapered helix or 2D spiral.
>
> **Enter Radius**
> This allows you to enter a constant radius in the Radius text box.
>
> **Turn Direction** determines the turn direction of the helix.
>
> **Right Hand** creates a helix in a counterclockwise direction.
>
> **Left Hand** creates a helix in a clockwise direction.

Define Orientation specifies the Z-Axis and X Point to define the helix orientation. If you do not specify an orientation, the current WCS is used.

Point Constructor defines the origin of the helix.

Exercise: Creating a helix

1. Create a part file **helix**.
2. Select **Insert | Curve | Helix**.
3. Enter values for the **Number of Turns**, **Pitch** and **Radius**.
4. Define the **Turn Direction** as **Right Hand** or **Left Hand**.
5. Click **Define Orientation**.

6. Pick a line or linear edge as the Z-Axis.
7. Pick a point to specify the helix orientation.
8. Define the base point of the helix. The base point is the center of the helix.
9. Click **OK** to create the helix.

Summary:

In this chapter, you learned how to:

- Create a New Sketch.
- Reattach a Sketch to a New Plane or Face.
- Create Sketch Curves.
- Change the Overall Preferences of the Sketcher.
- Mirror Sketch Geometry.
- Convert Sketch Geometry from Active to Reference and Vice Versa.
- Position Sketches Relative to Model Geometry.
- Change the Shape of a Sketch Using the Drag Option.
- Animate the Motion of a Sketch by Changing a Sketch Dimension Value Incrementally.
- Position Features to Centerlines and Curves.
- Edit the Defining String of a Swept Feature.
- Interrogate and Remove Existing Geometric Constraints.
- Create a Parabola
- Create a Hyperbola
- Create a Helix

Chapter 4

Solid Modeling

After completing this chapter, you will be able to:

- Create an Extruded Body.
- Create a Revolved Body.
- Understand the Benefits of Solid Modeling.
- Create Edge Blends.
- Create Cliff Edge Blends.
- Handle Blend Overflow Situations.
- Edit Blended Edges and Parameters.
- Chamfer Edges
- Offset Solid Faces.
- Hollow Bodies.
- Taper Solid Faces.
- Create Rectangular and Circular Instance Arrays.
- Mirror Bodies and Features.
- Trim Bodies.
- Create and Edit Thread Features.
- Recognize Design Intent, and Incorporate It into Solid Models.
- Recognize Parametric Modeling and Its Advantages.
- Know the Role of Features in Solid Modeling.
- Perform Basic Edits on Features.
- Use the Model Navigator.
- Set Modeling Preferences.
- Create Primitive Features, Such as Blocks, Cylinders, and Cones.
- Use Boolean Operations (Unite, Subtract, and Intersect) to Alter a Solid Body.
- Apply Transformations to Objects.
- Translate a Selected Object in Space.
- Rotate Objects about a Point or Line.
- Mirror Objects through a Line or Plane.
- Create a Rectangular or Circular Array of Objects.
- Reposition Objects.
- Rotate Objects Between Two Axes.
- Perform Boolean Operations with Swept Features.
- Create an Extruded Body by Specifying One or Two Trim Surfaces.
- Extrude a Feature through Multiple Bodies.
- Create a Revolved Body by Specifying One or Two Trim Surfaces.
- Sweep a Section along a Guide String.

- Create a Tube Feature.
- Edit Swept Features.

Swept Features:

Swept Features are extruded or revolved profiles used to create a sheet body or solid body feature. The Swept feature types, found on the Insert | Form Feature menu, are Extrude, Revolve, Sweep along Guide, and Tube.

Before creating a swept feature, you must select a profile or section string. Section Strings are comprised of curves, sketches, sheet bodies, or face edges and can define an enclosed (closed) profile. If Body Type is set to Solid (in Preferences | Modeling), any swept feature operation can create a solid body from an enclosed section string.

Section Strings can also form an un-enclosed (open) profile. When this is the case, the endpoints of the section strings highlight with an asterisk. If more than one curve is selected, the curves must be coplanar. When an open profile is swept without using any offsets, a sheet body is created. If it is swept with Offset options, a solid body can be created if it's Body Type is set to **Solid**.

The solid or sheet body created from a swept operation is associated to its section strings. The section strings are then known as "smartcurves." If a datum axis is selected for use as an axis or direction vector, it is also associated to the solidor sheet body.

There are five options available when defining a section string for a swept feature.

Solid Face:

This automatically selects all the edges of a solid face to use as the section string. If the selected face is modified, the resultant sweptfeature is also modified. Once used in the swept feature, the selected faces cannot be deleted because of their associativity.

Solid Edge:

This allows individual edges of solid faces to be selected. If the selected edge is modified, the resultant swept feature is also modified. Once used in the swept feature, the selected edges cannot be deleted because of their associativity.

Curve:

This allows the selection of curves or individual sketch curves. If the selected curves are modified, the resultant swept feature is also modified. Once used in the swept feature, the selected curves cannot be deleted because of their associativity.

Chain Curves:

This allows the selection of curves that are end-to-end contiguous. Unigraphics prompts for the selection of the Chain Start and then the Chain End. The direction of the chaining action is determined by the endpoint nearest the selection point of the Chain Start. Chain Curves selects all of the connected curves between the Chain Start and Chain End. The Chaining Tolerance (in Preferences | Selection) defines the maximum allowable gap size that the chaining operation can bridge. Any gap larger than the specified chaining tolerance stops the selection process at that point.

Sheet Body:

This allows you to sweep one or more sheet bodies into a single solid

Note: Can I mix object types when defining a section string?
Yes. However, to create a single body, all the selected objects (curves, edges, or faces) must still result in a contiguous profile.

Section strings for different features can be defined in the same sketch, if you select the curves for each feature individually using the Curve option.

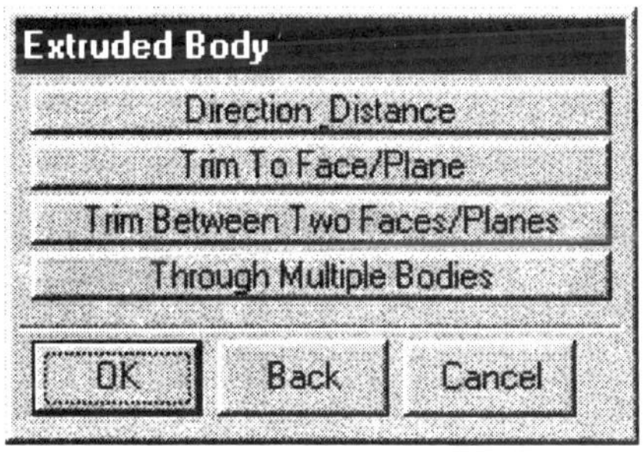

Extrude:

Extrude Using Direction and Distance:

 Create a body by sweeping a section string over a specified linear direction and distance.

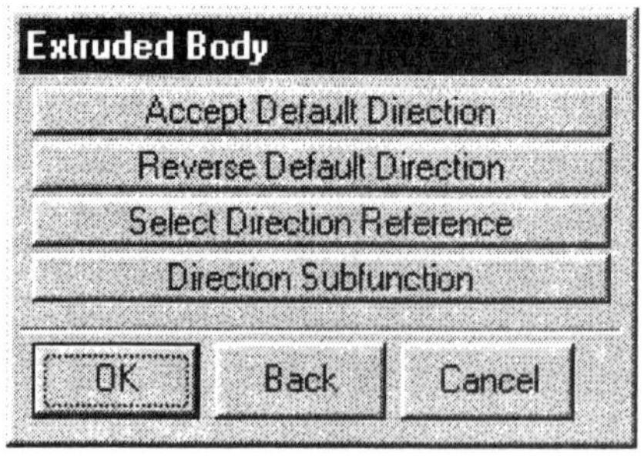

Direction and Distance:

This sweeps a section string in a specified direction for a specified distance. If the selected section string objects are coplanar, the default reference direction is normal to the plane of the section string. This can be reversed or redefined completely. If the section strings are not coplanar, the extrusion direction must be defined by choosing **Select Direction Reference** or **Direction Subfunction**.

> **Accept Default Direction** confirms the displayed direction vector.
>
> **Reverse Direction Vector** reverses the displayed direction vector.
>
> **Select Direction Reference** allows the selection of a linear Edge Endpoint, Datum Axis, or a Line End Point to define the direction vector.
>
> The endpoint nearest where the linear edge or line is selected determines the positive direction.
>
> **Direction Subfunction** uses the **Vector Constructor** dialog to define the direction vector.

Start and End distance:

These values determine the beginning and end of the extrusion. A value of zero for either the start or the end distance is in the same plane as the section string. Values may be positive or negative. The difference between the start and end values determines the overall length of the extrusion, which cannot equal zero.

First and Second Offset:

The Offset options allow the creation of a feature offset from the selected section string by a constant value. A value of zero for either the first or second offset defines the edge(s) of the resultant body as being coincident with the section string.

In the Graphics window a dashed vector arrow represents the positive offset direction. Both positive and negative values are acceptable for input as offset values. None, one, or both of the offsets may be specified. The offset direction is always perpendicular to the extrusion direction.

Closed Profile:

When the section string forms a closed profile, the offsets define an inner and outer 'wall' based on the section string. If no values are defined, a solid body is created from the section string.

Open Profile:

When the section string does not form a closed profile, the offsets determine the type of body created. If the difference between the first and second offset is greaterthan zero, a solid body is created with constant offsets from the section string. If the difference between the first and second offsets is zero, a sheet body is created.

Taper Angle:

This tapers faces of the extruded feature as they are created. Only planar, cylindrical, or conical bodies with an axis parallel to the selected extruded direction can be tapered as the extrusion is created. If there is no solution to the extrusion for the specified taper angle, an error message displays.

A "positive angle" causes the sides of the feature to taper inward, toward the centroid of the section strings. If the defined angle is too steep for the extrusion length, the end face is undefined. When this situation occurs, an error message displays.

A negative angle causes the sides of the feature to taper outward, away from the center of the section strings.

A "zero" taper angle results in no taper being applied.

Taper Start – At Start Distance:

Use this to define the taper as starting from the extrusion Start Distance.

Taper Start – At Defining String:

Use this to define the taper as starting at the section string.

Exercise: Extruding a section string by direction and distance

1. Open **seed_in**.
2. Application **Modeling**.
3. Select **Insert | Form Feature | Extrude**
4. The **Extruded Body** dialog displays prompting for an extrusion type.

5. Select **Direction and Distance**. An arrow displays in the Graphics window indicating the extrusion direction, and the **Extruded Body** dialog re-displays allowing you to accept or re-specify this direction.
6. Accept, reverse, or define the Direction Vector. The **Extruded Body** dialog (for feature parameters) displays.
7. Enter the **Start Distance**, and **End Distance**. If needed, enter the **First Offset**, **Second Offset**, **Taper Angle**, and **Taper Start** values. Click **OK** to display the **Boolean Operation** dialog.
8. Select the appropriate Boolean operation to create the extrusion.

Extrude; Trim to Face or Plane:

 **Insert | Form Feature | Extrude**

Filter:

Use this to limit the type of objects selectable as trimming faces. The available choices are **All**, **Solid Face**, or **Datum Plane**. The default selection is **All**.

Trimming Face Operations:

Do not Extend Trim Face, does not attempt to extend the selected trimming face. Utilize this if the trimming face intersects the tool solid. An error message displays if the tool solid does not intersect the trimming face.

Extend Trim Face attempts to extend the trimming face to achieve a successful result. The face is extended along its natural curvature. An error message displays if Unigraphics cannot extend the trimming face to intersect the tool solid.

Try Both Options is the best setting to use to achieve a successful operation. Unigraphics attempts the operation by first using the Do not Extend Trim Face option. If the operation results in an error, the Extend Trim Face is then attempted.

Exercise: Extruding a section string using trim to face/plane

1. Open **mod_1**.
2. Application **Modeling**.
3. Select **Insert | Form Feature | Extrude**
4. If needed, select a section string type, then pick the section string and click **OK**.
5. Select **Trim To Face/Plane**. An arrow displays in the Graphics window indicating the extrusion direction, and the **Extruded Body** dialog displays allowing you to accept or re-specify this direction.
6. Accept, reverse, or define the Direction Vector.
7. Select the necessary trim face options, pick the trim face, and click **OK**.
8. Select the appropriate Boolean operation to create the extrusion.

Notes: When trimming to a datum plane, the plane used must be derived from the target solid. Planes derived from fixed datum planes cannot be used for trimming. In addition, planes derived from bodies other than the target body cannot be used.

Create is not available when using Trim to Face. Only the Boolean operations involving another body (Unite, Subtract, and Intersect) can be used.
B-surface faces typically cannot extend very far when they have a high degree of curvature.

Extrude; Through Multiple Bodies:

 **Insert | Form Feature | Extrude**

Exercise: Extruding a section string using thru multiple bodies

1. Select **Insert | Form Feature | Extrude**
2. If needed, select a section string type, then pick the section string and click **OK**.
3. Select **Thru Multiple Bodies**.
4. The **Extruded Body** dialog re-displays prompting you to pick the target bodies.
5. Select the required target bodies, and click **OK**. The **Extruded Body** dialog (for feature parameters) displays.
6. If needed, enter the **First Offset** and **Second Offset**, then click **OK** to create the extrusion.

Notes: Only subtractions can be performed with **Thru Multiple Bodies**, so you are not prompted for a Boolean operation type.

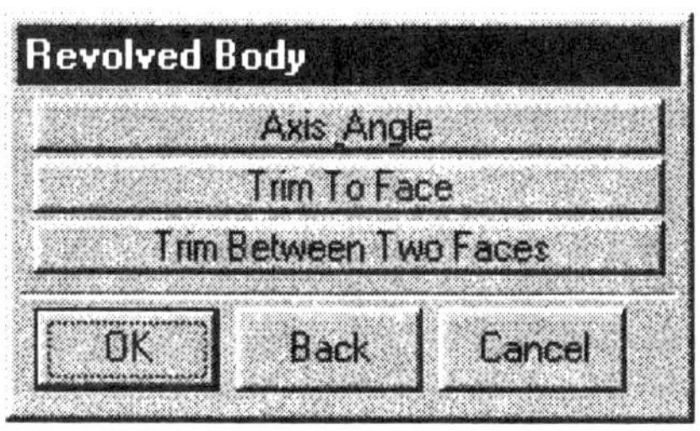

Revolve:

Body of Revolution; Axis and Angle:

 **Insert | Form Feature | Revolve**

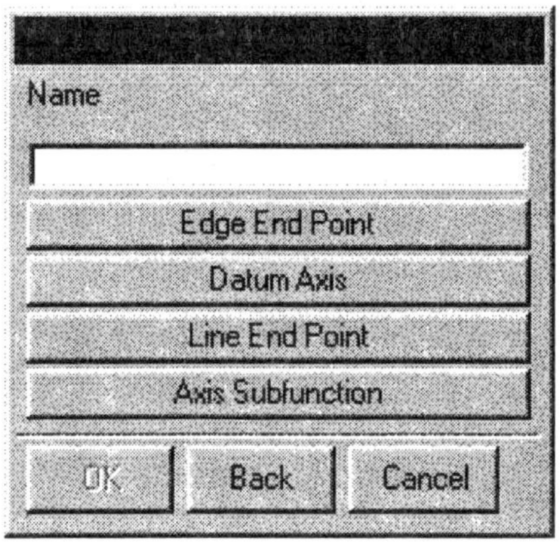

Edge End Points:

Pick a linear solid edge to use as an axis. The side selected on the designated edge midpoint defines the positive vector direction.

Datum Axes:

This allows you to pick an existing datum axis from the Graphics window. The positive vector direction is the same as that for the datum axis.

Line End Point:

Select a line to use as an axis. The side selected on the designated line midpoint defines the positive vector direction.

Axis Subfunction:

This uses the **Point Constructor** to define the origin point for the rotation axis. The **Vector Constructor** is then used to define a direction vector for the axis.

Start Angle and End Angle:

These define the beginning and end of the revolution. The plane of the section strings is considered zero degrees. The Start Angle is measured from zero, andthe End Angle is measured from zero in a counterclockwise direction. The total number of degrees must not exceed 360. The section string revolves around the axis from the start angle to the end angle. Positive or negative values can be entered.

First Offset and Second Offset:

These are identical to those used in **Extrude**.

The Offset allows the creation of a feature offset from the selected section string by a constant value. A value of zero for either the first or second offset defines the edge(s) of the resultant body as being coincident with the section string.

In the Graphics window, the positive offset direction displays as a *dashed* vector arrow. Both positive and negative values are acceptable for input as offset values. None, one, or both of the offsets can be specified. The offset direction is always perpendicular to the extrusion direction.

Closed Profile:

When the section string forms a closed profile, the offsets define an inner and outer 'wall' based on the section string. If no values are defined, a solid body is created from the section string.

Open Profile:

When the section string does not form a closed profile, the offsets determine the type of body created. If the difference between the first and second offset is greater than zero, a solid body is created with constant offsets from the section string. If the difference between the first and second offsets is zero, a sheet body is created.

Exercise: Revolving a section string using axis and angle

1. Open **seed_mm**.
2. Application **Modeling**.
3. Select **Insert | Form Feature | Revolve**
4. If needed, select a section string type, then pick the section string and click **OK**.
5. The **Revolved Body** dialog displays prompting for a revolve type.
6. Select **Axis _Angle**.
7. Pick or define the revolution axis, and click **OK** to display the **Parameters** dialog.
8. Input the **Start Angle** and **End Angle**. If needed, enter the **First Offset** and **Second Offset**. Click **OK** to display the **Boolean Operation** dialog.
9. Select the appropriate Boolean operation to create the revolution.

Notes: The rotation direction follows the right hand rule. For example, the direction is clockwise when looking along the axis in the direction of the arrow.
An open profile can form a solid body if the string is revolved between 0 and 360 degrees. Unigraphics will automatically cap the ends of the body.
Section strings must not cross the revolution axis, though they may be coincident with it.

Body of Revolution; Trim to Face:

 **Insert | Form Feature | Revolve**

Filter:

This can be used to limit the type of objects selectable as trimming faces. The available choices are **All**, **Solid Face**, or **Datum Plane**. The default selection is **All**.

Exercise: Revolving a section string using trim to face

1. Open **mod_3**.
2. Application **Modeling**.
3. Select **Insert | Form Feature | Revolve**
4. Pick the section string and click **OK**. The **Revolved Body** dialog displays prompting for a revolve type.
5. Select **Trim To Face**.
6. Pick the trim face and click **OK**. The reference axis dialog displays allowing you to define a revolution axis.
7. Pick or define the revolution axis.
8. If needed, enter the **First Offset** and **Second Offset**. Click **OK** to display the **Boolean Operation** dialog.
9. Select the appropriate Boolean operation to create the revolution.

Notes: Datum planes may be used in this operation, provided they are associated to the target body. Therefore, fixed datum planes may not be used.
Create is not available when using Trim to Face. Only the Boolean operations involving another body (Unite, Subtract, and Intersect) can be used.
The trimming face must completely intersect the tool solid.

Swept along a guide:

 **Insert | Form Feature | Sweep along Guide**

Section Strings:

The options for selecting the section string are the same as Extruded Body and Body of Revolution. Section strings do not have to be planar, but non-planar section strings do create a sheet body without any end caps or sides. The best results are obtained with planar section strings.

Guide Strings:

The guide string may be comprised of faces, curves, sketches or solid edges. More than one curve or edge can be used as a guide string. If this is the case, the curves or edges must be continuous, but not necessarily tangent. The guide string can contain sharp corners, and is not required to be planar.

Exercise: Sweeping a section string along a guide

1. Open **mod_4**.
2. Select **Insert | Form Feature | Sweep along Guide**.
3. Pick the section string and click **OK**. The **Sweep along Guide** dialog displays prompting for a guide string.
4. Pick the guide string and click **OK** to display the **Enter Offsets** dialog.
5. If needed, enter the **First Offset** and the **Second Offset** values and click **OK**. The **Boolean Operation** dialog displays.
6. Select the appropriate Boolean operation to create the sweep.

Notes: If the section string has multiple loops (for example, an inner ID and an outer ID swept in the same feature) the guide string must be comprised of lines and arcs.
This operation works best when the section string is planar, normal to the guide string and located at the starting point of the guide string. If the section string is not on the starting point of the guide string, undesirable results may be obtained.
Try to avoid self-intersecting conditions. If the guide curve has a radius too small for the size of the section string or if the guide string contains two adjacent lines that form an acute angle, a solution is difficult.

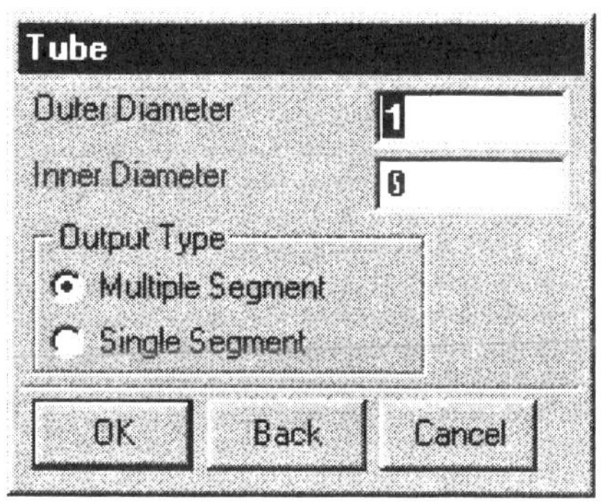

Tube:

 **Insert | Form Feature | Tube**

Outer Diameter:

This is the outer diameter of the tube. It cannot be zero.

Inner Diameter:

This is the inner diameter. A diameter of zero creates a solid or cable.

Output Type:

This controls the way UG constructs the tube feature. The Output Type cannot be changed during the editing function.

Multiple Segment, creates multi-segment tubes, consisting of cylindrical and toroidal faces, which are mathematically simpler. This produces a feature with multiple faces with a reduced file size as compared to the complex representation of a Single Segment tube.

Single Segment tubes are more complex representations made up of B-surfaces. The resulting tube face complexity is determined by the complexity of the guide string. A higher degree guide string produces a more complex tube face.

Guide Strings:

The guide string is the path along which the tube is swept. Tube features can be swept along entire sketches, partial sketches, regular curve geometry, solid edges or faces.

Exercise: Creating a tube

1. Open **seed_in**.
2. Application **Modeling**.
3. Select **Insert | Form Feature | Tube**.
4. Input values in the **Outer Diameter** and the **Inner Diameter** text boxes.
5. Select **Single** or **Multiple** segments. Click **OK** to display the **Tube** dialog for selecting the guide string.
6. Pick the guide string and click **OK**. The **Boolean Operation** dialog displays.
7. Select the appropriate Boolean operation to create the tube.

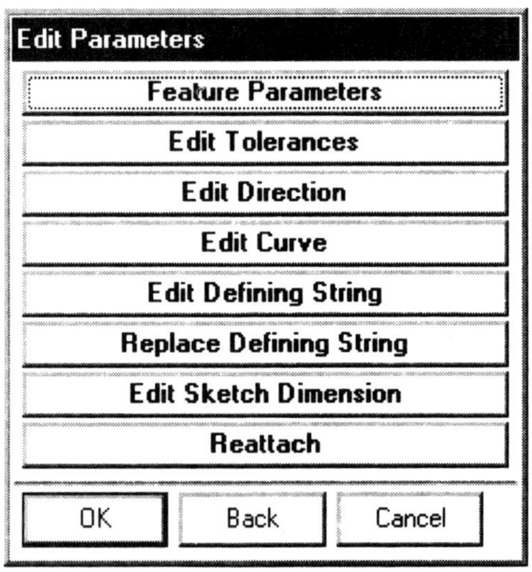

Editing Swept Features:

Swept features (Extruded Body, Body of Revolution, Sweep along Guide and Tube) allow for powerful and flexible editing after their creation. You can change the parameters used in the creation of the feature, and alter the section string, or entirely replace it.

There are two ways to edit a swept feature's parameters. The first is to select **Edit | Feature | Parameters** and pick the feature to edit from the dialog. The second is too right-click the feature in the Model Navigator and select Edit Parameters from the pop-up menu.

Extruded Body:

When **Edit | Feature | Parameters** is selected and an Extruded feature is selected, the Edit Parameters dialog displays listing the available edit selections.

Selecting Feature Parameters displays a dialog with the creation parameters for the feature.
You can change the Start Distance, End Distance, the Offsets (if applied), Taper Angle, and the Taper Start for an extrusion.

Selecting Edit Direction displays the Vector Constructor, allowing you to change the direction in which the extrusion is applied.

Body Revolution:

When **Edit | Feature | Parameters** is selected and a revolved feature is selected, the Edit Parameters dialog displays with the available edit selections.

Selecting Feature Parameters displays a dialog with the creation parameters for the feature.
You can change the Start Angle and End Angle. Offsets on Revolved features can only be edited if they were applied when the feature was created. You cannot add them later.

Swept along a Guide:

When **Edit | Feature | Parameters** is selected and a Sweep feature is selected, the Edit Parameters dialog displays with the available edit selections.
If no offsets were applied, Feature Parameters is not available. If offsets were applied, they can be edited by selecting this option.

Tube:

When **Edit | Feature | Parameters** is selected and a Cable (tube) feature is selected, the Edit Parameters dialog displays with the available edit selections.

Selecting **Feature Parameters** allows you to change the inner and outer diameters.

Edit Defining String(s):

This allows for editing of the Section String and/or the Guide String. Objects may be added to or removed from the strings.
Extruded, Revolved, and Sweep features allow the Section Strings to be modified. Cable and Sweep features allow Guide Strings to be modified.

To remove objects from the string, press the SHIFT key down and pick the objects from the Graphics window. To add new objects to the string, pick a curve, edge, or face to add to the definition from the Graphics window.

Replace Defining String:

This allows for the removal and replacement of current section or guide strings.
This replaces a contiguous string of curves in a string with a new set of contiguous strings. To remove the curves press the SHIFT key down and pick them. To add new curves, pick them from the Graphics window.

Edit Sketch Dimension:

If a feature was created using a sketch with dimensions an option to edit the sketch dimension appears in the Edit Parameters dialog for that feature.

Selecting **Edit Sketch Dimensions** displays a dialog with an expression list of dimensions for the sketch used to create the feature.
Select an expression and modify the expression name and/or the value. The dimensions also display in the Graphics window.

When **Delay Evaluation** is deselected, modifications to the expressions are instantly applied. When selected, modifications to the expressionsare delayed until **Apply** or **OK** are selected.

Alternate Solution is identical to that found in the Sketcher dialog. The system finds another solution to the selected dimensional constraint.

If an extracted curve is being utilized in the sketch, **Redefine Extracted Curve** provides an opportunity to select a different curve to extract.

Reattach:

This is available if the swept feature is created from a sketch. All sketches are attached to a planar face or datum plane. This allows the sketch to be relocated toa new planar face or datum plane.

The option available varies depending on the way the swept feature is created and from what objects the sketch is comprised.

Datum Features:

Datum features are construction tools that help you define orientations and locations for the placement of features. Datum features help define the function of the part or feature, manufacturing processes, methods of inspection, shape of the part, relationship to other features, assembly considerations and overall design requirements.

Datum Planes and Datum Axes:

Datum planes and datum axes are theoretical planes (any plane can be defined by three points) and axes (defined by any two points) created in Unigraphics either relative to the WCS or by defining constraints. Constraints provide the orientation of the plane or axis location relative to existing part geometry.

Datum Planes can be defined to be constrained to a solid object and remain associated with the model while it is being modified. You can use these planes as construction tools for creating and positioning features later in the model. When you modify a feature, the datum planes update because of this associativity. Any later features that reference these datum planes update their orientation or location relative to the datum planes. A Datum Axis is another construction tool that maintains associativity to the solid model.

Together, datum planes and datum axes are interchangeably referred to as reference features or datum features in Unigraphics.

Creating Associative Datum features:

When creating associative datum features, Unigraphics prompts you to select solid edges, faces or points on an existing solid. After the datum feature is created, it remains associated to the edges, faces or points of the solid unless youremove the edges, faces or points. This is because the faces/edges/points you select are constraints Unigraphics uses to define the plane or axis.

Fixed Datum Planes:

Unigraphics also allows you to create fixed datum features by specifying planes or axes relative to the WCS. You can also define a fixed datum plane using Plane Subfunction techniques and fixed datum axis using Vector Subfunction techniques. These planes and axis are considered fixed in space and are not associated to the existing model geometry.

When you click the **Datum Plane** icon, the **Datum Plane** dialog displays.

The fixed datum plane area at the bottom of the dialog contains the following options:**3 Planes of WCS**, **XC-YC**, or **YC-ZC**, or **XC-ZC** and **Plane Subfunction**.

By selecting the type from the list and clicking **OK** or **Apply** you create a fixed datum plane.

Notes: A good use for a datum plane is to add a form feature to a non-planar face. To accomplish this you must create a datum plane and attach the feature to the datum plane. An excellent example is when you are trying to place a hole into a cylindrical face, as shown in the figure.

If you delete a datum, any dependent features of the datum are also deleted. For example, if a datum plane is the placement face for a hole and the datum is deleted, the hole is also deleted. If you move the datum instead of deleting it, dependent features also move. In the previous example, if the datum is moved away from the solid, the depth of the hole may decrease. Keep in mind that datum planes and axesare features and parent-child relationships can exist

Datum planes and datum axes are infinite in size, the display that you see in the Graphics window serves only as a reference (and to allow for the selection of the feature). Their display is always just a bit bigger than the solid model geometry currently displaying.

Datum Planes:

This creates a reference plane feature by defining the plane relative to existing solid faces, endpoints, midpoints, edges, existing datum axes and existing datum planes, or relative to the WCS.

Datum Planes are Particularly Useful for:

- When selecting a planar placement face for a form feature or sketch.
- For defining a horizontal or vertical reference when creating form features and sketches.
- When creating positioning dimensions for form features and sketches.
- When mating components into an assembly.

Datum Plane Constraints:

The process of defining the plane is called **constraining**. In some situations, one constraint is sufficient to define a datum plane. For example, youcan create a datum plane that is offset from an existing datum plane or solid planar face. Because an existing datum plane or solid planar face already defines a plane, the offset distance is all that is required to define a new plane.
In other cases more than one constraint is required. For example, in the figure a datum plane is created in order to place a hole into a cylindrical face. In this case, a relative datum plane has to be created at the correct angle. A new relative datum plane is then created that is parallel to the first datum plane and tangent to the cylindrical face.

Notes: Use caution when working with fixed datums. If you position anything relative to a fixed datum, you must make sure that any edits to the body do not result in any Boolean errors. The fixed datum is fixed at a location in space.
If you pick a datum plane to define a horizontal or vertical reference, the intersection of the plane and the planar placement face defines the horizontal or vertical direction.

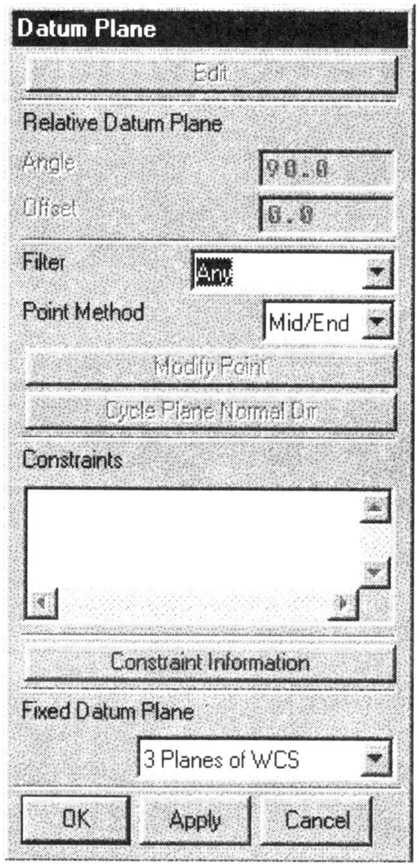

Create Datum Planes:

Insert | Form Feature | Datum Plane

Exercise: Creating an offset datum plane

1. Open **mod_5**.
2. Application **Modeling**.
3. Select **Insert | Form Feature | Datum Plane**.
4. Select a planar solid face or an existing datum plane.
5. Note **Offset to plane** displays in the **Constraints** list.
6. Enter a value in the **Offset** text box. A direction arrow indicates the positive offset direction.
7. Select **OK** or **Apply** to create the datum plane.

Exercise: Creating a datum plane at an angle to a face

1. Continue working with **mod_5**.
2. Select **Insert | Form Feature | Datum Plane**.
3. Select a solid edge to serve as the axis of revolution.
4. Select a planar solid to reference the angle from. The face does not have to be adjacent to the edge.
5. Note **Through edge** and **Angle to plane** displays in the **Constraints** list.
6. Enter a value in the **Angle** text box. A direction arrow indicates the positive rotation direction.
7. Select **OK** or **Apply** to create the datum plane.

Exercise: Creating a datum plane through the axis of a cylindrical face

1. Open **mod_6**.
2. Select **Insert | Form Feature | Datum Plane**.
3. Select a cylindrical or conical solid face.
4. Note **Through face axis** displays in the **Constraints** list.
5. Select **OK** or **Apply** to create the datum plane.

Exercise: Creating a datum plane tangent to a face

1. Continue with **mod_6**.
2. Select **Insert | Form Feature | Datum Plane**.
3. Create a datum plane through the axis of the cylinder.
4. Click **Apply** to create the datum plane.
5. Select the datum plane through the axis of the cylindrical face.
6. Select the cylindrical face. Five direction arrows appear,one dashed arrow in the axis direction, and four solid arrows normal to the cylindrical face.
7. Select one of the four direction arrows that point normal to the cylindrical face. The new datum plane is created at the location of the selected arrow.
8. Note **Normal to plane** or **Parallel to Plane** and **Tangent to face** display in the **Constraints** list.
9. Select **OK** or **Apply** to create the datum plane.

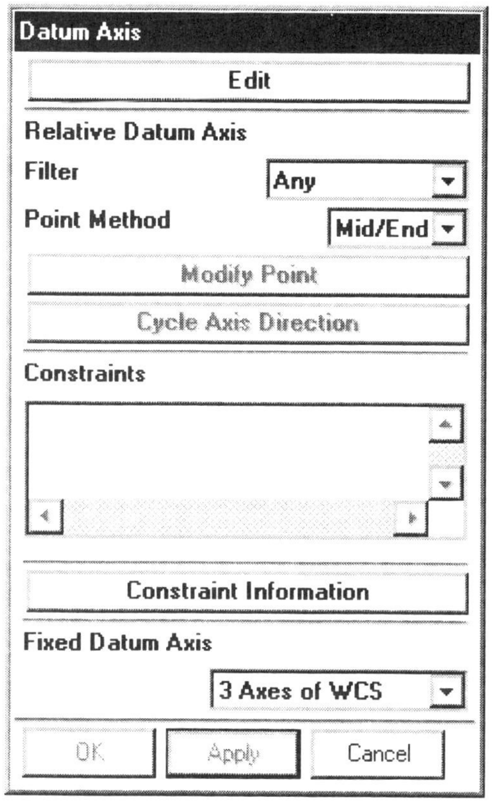

Datum Axis:

Insert | Form Feature | Datum Axis

Exercise: Creating a datum axis through the center of a cylinder

1. Continue with **mod_6**.
2. Select **Insert** | **Form Feature** | **Datum Axis**.
3. Select the cylindrical or conical face of a solid or sheet body.
4. **Through axis of face** displays in the **Constraints** list.
5. Select **OK** or **Apply**.

Exercise: Creating a datum axis through the intersection of two planes

1. Open **mod_5**.
2. Select **Insert** | **Form Feature** | **Datum Axis**
3. Select a datum plane or a planar solid face.
4. Select another datum plane or planar solid face. Selecting the faces in a counter-clockwise direction controls the datum axis direction.
5. Note two **Intersection of plane** constraints display in the **Constraints** list.
6. Select **OK** or **Apply**.

Notes: To pick a datum plane, pick one of the edges. To pick a datum axis, pick one of the endpoints.

Form Features:

Form Features are detail features that automatically unite or subtract from a solid body. Because they use relative positioning, they provide greater flexibility than primitives. Relative positioning allows you to parametrically position form features on the solid relative to other solid features. You can access the six form features, hole, slot, groove, pocket, boss, and pad, by selecting Insert | Form Feature.

Planer Placement Face:

All form features, with the exception of groove, require a planar placement face. The planar placement face is a flat surface on which to place the form feature. After defining the form feature parameters, you can position features on the planar placement face using relative positioning dimensions and you can move features to new locations on the planar placement face by modifying the relative positioning dimensions. Select **Reattach** (**Edit | Feature | Parameters**) to reposition form feature onto another planar placement face.

Horizontal Reference:

After selecting a planar placement face, the **Horizontal Reference** dialog displays prompting you for a horizontal reference.

The horizontal reference defines the XC or length direction for a non-cylindrical feature (rectangular pocket, pad and slot). Unigraphics requires horizontal and vertical directions, which are independent of the WCS, on the planar placement face.

A horizontal reference can be any solid edge, endpoint, datum axis or datum plane that defines a vector parallel to the planar placement face.

After assigning the horizontal reference, the vertical reference is assumed perpendicular to the horizontal reference on the planar placement face.

Notes: Can a hole or boss be placed on a non-planar face?
No, but they can be placed on datum planes. Locate a datum plane tangent to the curved face, and then place the hole on the datum plane.

You can assign a vertical reference instead of a horizontal. After you define the vertical reference, Unigraphics assumes the horizontal and then displays it.

Relative Positioning:

Relative positioning is the distance relationship of a feature (or a sketch) with respect to an existing solid edge, curve or datum plane.

Relative positions parametrically locate form features and sketches on a model. For example, a hole center may be a certain distance from one edge of a planar face and another distance from a perpendicular edge of the same face. If the planar face changes, the position of the hole maintains its specified distance. You can change the distances at any time. When features are dimensioned (i.e., holes or pads) on a drawing sheet their positions are dimensioned relative to some datum. Relative Positioning is the same idea.

When you add a feature to a solid, Unigraphics does not require that you fully specify the feature's position. The creation of a form feature can be complete without fully defining its location. By clicking **OK** before specifying a dimension type, Unigraphics automatically locates the feature at the same point where you picked the planar placement face.

If a horizontal reference has not been specified and you select a Horizontal or Vertical positioning dimension, you must select a reference.

The WCS is oriented to the horizontal reference when applying positioning dimensions. It returns to its original orientation and location after exiting the dialog. In addition, after assigning a horizontal reference, the WCS relocates to the new location and the horizontal dimension is the XC direction.

Positioning Options:

When positioning form features, the form feature being positioned is the **tool solid** and the solid on which it is being placed is the **target solid**.

 This assigns a positional constraint parallel to the horizontal reference or 90 degrees to the vertical reference. Because this is a point-to-point constraint, you cannot use datum planes.

 This assigns a positional constraint 90 degrees to the horizontal reference or aligned parallel to the vertical reference. Like **Horizontal**, this constrains one point relative to another; therefore you cannot use datum planes.

 This assigns a parallel positional constraint between two points, using entity endpoints, arc center points or arc tangent points. The dimension line is always parallel to the imaginary line created between the two points selected.

 This assigns a positional constraint perpendicular to a linear edge of the target solid to a point on the sketch or tool solid. You can also use datum planes, datum axis or an existing curve to define the target edge.

 This assigns a positional constraint to be parallel from a linear edge of a target solid to a linear edge of a sketch or tool solid.

 This assigns an angular positional constraint between a linear edge of the target solid or reference datum and a linear edge of the feature or sketch.

 This assigns a constraint dimension the same as Parallel but with the fixed distance between the two points set to zero, it makes the two points coincident.

 This assigns a constraint dimension the same as Perpendicular but with the fixed perpendicular distance between the edge and point set to zero.

 This assigns a constraint dimension the same as Parallel at a Distance but with the distance between the linear edge of the feature or sketch and the edge on the target solid set to zero (i.e., it makes the two edges or lines collinear).

Positioning a Feature to a Centerline:

Unigraphics can position slots, grooves, and rectangular pads and pockets to their centerlines, as well as to their linear edges. These centerlines display only when you initially position the feature and while editing the positioning dimensions; however, Unigraphics considers them a part of the feature and maintains an associative relationship with the feature to which they belong. When you reattach or modify a feature the centerlines automatically update, and if you delete a feature, the centerlines delete as well.

Slots, rectangular pads and rectangular pockets display centerlines running both horizontally and vertically.

Grooves maintain a centerline located an equal distance from both the top and bottom edges of the groove with a diameter equal to that of the groove. You can use any of the linear edge positioning methods to locate the feature.

Positioning Features to Curves:

You can position features relative to wire frame geometry and maintain complete associativity between the feature and the wire frame. This function is helpful because it provides flexibility when creating designs without having all of the geometric considerations completely defined or known.

When using this option, you can pick any curve type for positioning. For example, when using a perpendicular dimension, you must pick a line for the target edge. You can create curves after you create the feature and then reposition the feature to these curves.

Notes: If I locate a feature with one type of positioning dimension, can I later change to a different dimension type? Yes, by deleting the dimension and then adding the new dimension type.

Use Parallel to create a radial dimension.

Use Perpendicular to locate a cylindrical feature relative to a linear edge.

Use Parallel at a Distance to locate a linear edge relative to another linear edge.

Use Point onto Point to locate cylindrical features concentrically (i.e., center point to center point).

When placing a circular form feature, it is not necessaryto assign a horizontal reference if you don't use the Horizontal or Vertical relative positioning options to locate the feature. This is applicable for holes, cylindrical pockets and bosses.

When positioning features to curves, you cannot delete the curves used to define the target edges without deleting the associated positioning dimensions or feature.

Creating a Boss:

 **Insert | Form Feature | Boss**

Diameter sets the boss diameter measured at the base of the feature if taper is applied.

Height sets the height of the boss measured from the planar placement face to the top of the boss.

Taper Angle sets the taper angle (or draft) along the sides of the boss. An angle of 0 creates a boss with straight vertical sides. You can use a negative taper angle.

Exercise: Creating a boss

1. Open **mod_7**.
2. Select **Insert | Form Feature | Boss**.
3. Pick the planar placement face. The **Boss** dialog displays.
4. Type in the feature parameters and click **OK**. The **Positioning** dialog displays.
5. Set up the relative positioning dimensions and click **OK**.

Note: When creating a boss on a cylindrical or conical face, use a datum plane to specify a planar placement face.

Creating a Hole:

 **Insert | Form Feature | Hole**

Simple Hole creates a hole at a specified diameter. In the **Simple Hole** dialog enter values in the **Diameter**, **Hole Depth**, and **Tip Angle** text boxes.

If **Thru Hole** is selected, **Depth** and **Tip Angle** are not included in the dialog.

Counterbore Hole creates a hole with a counterbore.

In the **Counterbore Hole** dialog, enter values in the **Counterbore Diameter**, **Counterbore Depth**, **Hole Diameter**, **and Hole Depth** and **Tip Angle** text boxes. If **Thru Hole** is selected, **Hole Depth** and **Tip Angle** are not included in the dialog.

The hole depth must be greater than the counterbore depth and the counterbore diameter must be greater than the hole diameter.

Countersunk Hole creates a hole with a countersink.

In the **Countersink Hole** dialog enter values in the **C-Sink Diameter**, **C-Sink Angle**, **Hole Diameter**, **Hole Depth** and **Tip Angle** text boxes. If **Thru Hole** is selected, **Hole Depth** and **Tip Angle** are not included in the dialog.

The countersink angle must be greater than 0 and less than 180 degrees and the countersink diameter must be greater than the hole diameter.

Thru Hole will have the hole go from one face completely through the solid to another face without specifying a depth. Deactivate this to have the hole bore through one face to a specific depth. If the distances between the faces change, the hole remains associative to its through face.

Exercise: Creating a hole

1. Continue with **mod_7**.
2. Select **Insert | Form Feature | Hole.**
3. Click **Thru Hole** to make the hole go through another face on the solid, if necessary.
4. Select a hole type (**Simple**, **Counterbore**, **Countersink**).
5. Click **OK**. The relevant **Hole** dialog displays prompting you for a planar placement face.
6. Pick the placement face. If **Thru Hole** was selected, pick a face for the hole to go through. The relevant **Hole** dialog for setting parameters displays.
7. Type in the hole parameters and click **OK**. The **Positioning** dialog displays.
8. Set up the relative positioning dimensions and click **OK**. To place the hole without any positioning dimensions, click **OK**.

Note: You must select Thru Hole before selecting the hole type.

Creating a Rectangular Pad:

 **Insert | Form Feature | Pad**

Length sets the length of the pad, measured parallel to the horizontal reference.

Width sets the width of the pad, measured perpendicular to the horizontal reference.

Height sets the height of the pad, measured from the planar placement face to the top of the pad.

Corner Radius creates a fillet radius value for all four vertical edges of the pad. A value of zero produces sharp corners.

Taper Angle creates a taper on all four walls of the pad. The taper direction starts at the planar placement face and continues to the top of the pad.

Exercise: Creating a rectangular pad

1. Continue with **mod_7**.
2. Select **Insert | Form Feature | Pad.**
3. Click **Rectangular**. The **Rectangular Pad** dialog displays prompting you for a planar placement face.
4. Pick the placement face. The **Horizontal Reference** dialog displays.
5. Pick a horizontal reference. The **Rectangular Pad** dialog for setting feature parameters displays.
6. Type in the feature parameters and click **OK**. The **Positioning** dialog displays.
7. Set up the relative positioning dimensions and click **OK**. To place the pad without any positioning dimensions, click **OK**.

Cylindrical Pocket:

 **Insert | Form Feature | Pocket**

Pocket Diameter sets the diameter of the pocket, measured at the planar placement face location.

Depth sets the depth of the pocket, measured from the planar placement face to the bottom of the pocket.

Floor Radius adds a radius between the cylindrical wall of the pocket and the pocket floor.

Taper Angle adds a taper angle to the cylindrical wall of the pocket. The taper direction starts at the planar placement face and continues to the bottom of the pocket.

Exercise: Creating a cylindrical pocket

1. Continue with **mod_7**.
2. Select **Insert | Form Feature | Pocket**.
3. Click **Cylindrical**. The **Cylindrical Pocket** dialog displays.
4. Pick the placement face. The **Cylindrical Pocket** dialog displays prompting you to enter pocket parameters.
5. Enter the parameters and click **OK**.
6. Set up the relative positioning dimensions and click **OK**. To place the pocket without any positioning dimensions, click **OK**.

Notes: Why not just use a simple hole? A cylindrical pocket is better than a simple hole for making holes in injection molded parts (because the draft angle can be specified in the feature itself), or if you need a floor radius (for a hole machined with an end mill).

Use a cylindrical pocket instead of a hole to position a linear edge to the end point or tangent point of the cylindrical pocket. Positioning dimensions of a hole feature are always to the arc center.

Rectangular Pocket:

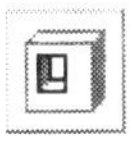 **Insert | Form Feature | Pocket**

X Length sets the length of the pocket, measured in the X direction, parallel to the horizontal reference.

Y Length sets the width of the pocket, measured perpendicular to the horizontal reference.

Z Length sets the depth of the pocket, measured from the planar placement face to the pocket floor.

Corner Radius creates a fillet radius value for all four vertical edges of the pocket. A zero value produces sharp corners.

Floor Radius adds a radius between the four vertical walls of the pocket and the pocket floor.

Taper Angle creates a taper on all four walls of the pocket. The taper direction starts at the planar placement face and continues to the pocket floor.

Exercise: Creating a rectangular pocket

1. Continue with **mod_7**.
2. Select **Insert | Form Feature | Pocket**.
3. Click **Rectangular**. The **Rectangular Pocket** dialog displays.
4. Pick the placement face. The **Horizontal Reference** dialog displays.
5. Pick a horizontal reference. The **Rectangular Pocket** dialog displays prompting you to enter the pocket parameters. A direction vector displays indicating the horizontal reference, which corresponds to the **X Length** parameter.
6. Enter the feature parameters and click **OK**. The **Positioning** dialog displays.
7. Set up the relative positioning dimensions and click **OK**. To place the pocket without any positioning dimensions, click **OK**.

Slots:

 **Insert | Form Feature | Slot**

Thru Slot works differently than **Thru Hole** in that a thru slot is created through two faces that are not coplanar to the slot placement face.

When this is activated, **Length** does not appear on the various **Slot** dialogs.

Rectangular creates a slot with sharp corners between its sides and floor.

Ball End creates a slot with a full radius at the bottom of the slot.

U – Slot creates a slot with corner radii in the bottom of the slot

T – Slot creates an inverted "T" shaped slot.

Dove – Tail creates a slot with angled sides.

Exercise: Creating a slot

1. Continue with **mod_7**.
2. Select **Insert | Form Feature | Slot**.
3. Activate **Thru Slot** to make the slot go through two faces on the solid. Select a slot type and click **OK**. The relevant **Slot** dialog displays prompting you for a planar placement face.
4. Pick the placement face.
5. Pick a horizontal reference. If you are using the **Thru Slot** feature, pick both faces for the slot to pass through. Otherwise, the relevant **Slot** dialog for setting parameters displays.
6. Type in the hole parameters and click **OK**.
7. Set up the relative positioning dimensions and click **OK**. To place the hole without any positioning dimensions, click **OK**.

Groove:

 **Insert | Form Feature | Groove**

Rectangular creates a groove with sharp corners in the bottom of the groove.

Ball – End creates a groove with a full radius in the bottom of the groove.

U – Groove creates a groove with corner radii in the bottom of the groove.

Relative Positioning of Grooves:

The groove feature uses a different method for defining its position than the other form features.

After you enter the groove parameters, Unigraphics prompts you to select a target edge. Pick an edge on the existing solid to establish the first location of the dimension. Next you are prompted to pick a tool edge. Pick an edge on the groove to establish the second location of the dimension. Enter the appropriate values for the location of the groove. This is the only positioning dimension needed for the groove because the groove is centered on the axis of the cylindrical or conical placement face.

Exercise: Creating a groove

1. Open **mod_8**.
2. Select **Insert | Form Feature | Groove**.
3. Select a groove type and click **OK**. The relevant **Groove** dialog displays prompting you for a placement face.
4. Pick a cylindrical, conical or spherical placement face. The relevant **Groove** dialog for entering feature parameters displays.
5. Type in the feature parameters and click **OK**. The **Position Groove** dialog displays.
6. To place the groove without a positioning dimension, click **OK**. Otherwise, pick an edge on the target body, then pick an edge on the tool body. The **Create Expression** dialog displays prompting you for a distance between the two edges.
7. Enter a value and click **OK**.

Notes: Grooves can be internal or external as shown in the following figure.
Positioning dimensions can reference the center of a groove, in addition to either edge.

Editing Form Features:

You can edit form features two different ways: by editing their creation parameters (size and shape) and by editing their positioning dimensions. It is also possible to reattach form features by moving them to a new face or datum plane.

 Selecting **Parameters** displays a dialog that allows you to select individual features for editing. You can select features from the list, pick them in the Graphics window, or select them from the Model Navigator by right clicking on them.

After selecting a feature, a dialog displays allowing you to edit the feature parameters or reattach the feature.

Selecting **Feature Dialog** displays the same dialog that was used to enter the feature's initial parameters. You can change these parameters as neededand the feature updates accordingly.

Reattaching form Features:

Selecting **Reattach** from the **Edit Parameters** dialog displays the **Reattach** dialog. From here you can change the placement face of the form feature or change the horizontal reference (if applicable). Other options are available for certain features, for example, redefining the through face.

 Selecting **Positioning** displays a dialog that allows you to select individual features for repositioning. You can select features from the list or pick them in the Graphics window.

If the feature was placed without any positioning dimensions, then the**Positioning** dialog displays allowing you to add dimensions. If the feature had dimensions added to it when it was created, a dialog displays allowing you to edit those dimensions. You can edit existing dimension values, delete existing dimensions or add new dimensions.

Notes: Can the form feature type change after it has been created? You can change the type within a feature family. For example, you could change a Rectangular Slot to a Dove-Tail Slot. However, you cannot change families.

If you have a large number of features to edit, especially in a complex model, select**Edit | Feature | Parameters**. This allows you to edit several features at once without causing the entire part to rebuild. Right clicking in the Model Navigator and selecting Edit Parameters causes the part to update after you finish each feature.

Edge Blending:

Edge blends add or remove material on a solid body by replacing selected edge(s) with cylindrical or conical surfaces. Edge blends are created using**Insert | Feature Operation | Edge Blend** and can have either a constant or variable radius.

There are two types of edge blends: **Constant Radius** and **Variable Radius**. A variable radius blend can be defined by specifying different radii at different points along the selected edge(s). The radius of the blend is interpolated between the specified points.

 **Insert | Feature Operation | Edge Blend**

Filter filters the edge selection. Although blends apply to edges of solids, you can pick all of the edges in a face or body at one time by specifying the appropriate filter. The default **All** allows you to pick individual edges, all edges in a face or all edges in a body based on the **Accept/Next Object** dialog. This drop-down list is not available when the blend type is set to **Cliff Edge**.

Default Radius allows you specify the blend radius.

Add Tangent Edges will select all of the edges that are tangent to a selected edge.

Blend All Instances to blend one feature and then have all of the instance features blend automatically.

Use **Point Method** to choose between **Infer** options or the **Point Constructor** to pick points for a variable radius blend. The points picked and the defined radius display inthe list box.

Modify Point will allow you to adjust the location of a point selected for a variable radius blend. To use this option, select a point from the list box and click**Modify Point** to display the **Modify Point** dialog.

The **Modify Point** dialog contains two options, **% Arc length** and **Arc length**. You can modify the values by moving the slider bar or by typing a value in the text box next to the selected option. Click **OK** to lock the point at the chosen location.

Remove Point removes points from a variable radius blend. Removing all points converts the blend to a constant radius blend with the radius determined by the**Default Radius** value.

Exercise: Creating a constant radius blend

1. Open **mod_9**.
2. Select **Insert | Feature Operation | Edge Blend**.
3. Pick the edges to blend.
4. Enter a value in the **Default Radius** text box.
5. Click **Apply** or **OK**.

Exercise: Creating a variable radius blend

1. Open **mod_10**.
2. Select **Insert | Feature Operation | Edge Blend**.
3. Pick the edges to blend.
4. Select a **Point Method**. Note that you can select **Cursor Location** and the point you pick is projected onto the edge(s).
5. Pick a point. The dialog changes, enabling variable radius options. The point and its corresponding radius value appear in the list.

6. Enter a radius value for the point in the **Variable Radius** text box. The list updates showing the new value.
7. Continue adding points and point radius values. The list box in the dialog displays the points picked, along with the radius for each point.
8. If you pick an edge not needed, click **Respecify Edge** and pick a different edge.
9. To change the position of a point along an edge, click **Modify Point**. The **Modify Point** dialog displays, allowing you to move the point along the edge by entering a value in the text box or using the slider.
10. After specifying all desired points and point radii, click **OK** or **Apply**.

Exercise: Creating a cliff edge blend

1. Open **mod_11**.
2. Select **Insert | Feature Operation | Edge Blend**.
3. Pick the edges to blend.
4. Set the Blend Type to **Cliff Edge**.
5. Pick the blend edge(s).
6. Pick the cliff edge(s).
7. Enter a radius in the **Default Radius** text box.
8. Click **Apply** or **OK**.

Notes: If I create a constant radius blend, can I change it into a variable radius blend? Yes, select **Edit | Feature | Parameters** and add the points. You can also convert a variable radius blend to a constant radius by removing the points.

While picking edges to blend, you can deselect an edge by pressing SHIFT and picking the edge.
Explicitly defining a cliff edge blend is usually necessary for complex geometry.
When creating a variable radius blend with a value of zero at each end, you must specify at least one other point in the middle.

Overflow Blends:

Unigraphics creates an edge blend unless a blend interferes with other edges, faces or blends.

To help illustrate this concept, imagine a sphere with the same radius as the blend being rolled along the edges of the solid. Where the sphere is tangent to the faces, the material adjacent to the edges is either added or removed. Where the sphere contacts another face or edge outside of the adjacent faces, an overflow condition results.

An overflow can result from blends interfering with one another or blends interfering with other faces and/or edges.

Overflow Options:

When you click **Overflow Options** in the **Edge Blend** dialog, the following dialog displays.

Allow Smooth Overflows:
A smooth overflow occurs when the region of the overflow is smooth, typically when another blended face is in the area of the overflow.

Allow Cliff Edge Overflows:
When a cliff edge overflow occurs, Unigraphics converts the blend overflow to a cliff edge blend and the blend is "trimmed" where the overflow occurs. This results in a blend that does not overflow and has sharp edges in the overflow region.

Allow Notch Overflows:
When a notch overflow occurs, the blend remains the same as if it was created before the features that caused the overflow condition. The features that caused the overflow trim the blend.

You can control the types of overflow options used or let Unigraphics determine the appropriate overflow condition automatically. Unigraphics determines the overflow condition by default. If Unigraphics detects an overflow, it attempts to create a smooth overflow first. If the blend cannot be created, a cliff edge overflow is then attempted. If the blend still cannot be created, a notch overflow is attempted. If Unigraphics cannot generate any blend, the blend fails.

If you want to control the overflow options, select the overflow options you need. You cannot control the order in which they are applied. After an overflow blend is created, you can edit the overflow options by selecting **Edit | Feature | Edit Parameters**.

Note: It is usually best to use the default overflow conditions since this generally creates the smoothest blend possible.

Editing Blends:

 **Edit | Feature | Parameters**

After choosing a blend feature to edit, you can add or remove edges from the string by picking edges in the Graphics window. Deselect edges by pressing SHIFT and picking the edge.

Exercise: Editing a constant radius blend

1. Open **mod_12**.
2. Select **Edit | Feature | Parameters**.

3. Pick the blend in the Graphics window or from the **Edit Parameters** list box and click **OK**.
4. Enter a new value in the **Default Radius** text box.
5. Add or remove edges from the blend if necessary.
6. Click **OK**.

Exercise: Editing a variable radius blend

1. Continue with **mod_12**.
2. Select **Edit | Feature | Parameters**.
3. Pick the blend in the Graphics window or from the **Edit Parameters** list box and click **OK**.
4. Pick a point at the location where you want to change the radius.
5. Pick a new point along the string to add a point to the variable radius.
6. Click **Modify Point** to change the location of a variable radius point.
7. Click **Remove Point** to remove a point from the variable radius.
8. Click **OK**.

Notes: Do I have to suppress a blend before I can remove edges from it?
No, you can deselect blend edges without suppressing the blend.

You can add points to a fixed radius blend in order to create a variable radius blend. You can also remove all points from a variable radius blend to create a fixed radius blend.

Chamfers:

 **Insert | Feature Operation | Chamfer**

Single Offset creates a 45° chamfer by applying an offset value along each face or sheet forming the edge.

Double Offset allows you to enter both the first and second offset values. Unigraphics applies these values along all of the faces/sheets forming the selected edge(s).

Offset Angle creates a chamfer with a user-specified offset and angle.

Flip Last Chamfer is used when creating the **Double Offset** or the **Offset Angle** option; click **Flip Last Chamfer** in the **Chamfer** dialog to change the direction of the chamfer.

Exercise: Creating a single offset chamfer

1. Open **mod_7**.
2. Select **Insert | Feature Operation | Chamfer**.
3. Click **Single Offset**.

4. Pick the edge(s) to chamfer and click **OK**.
5. Enter a value in thc **Offset** text box.
6. Click **OK**.

Exercise: Creating a double offset chamfer

1. Continue with **mod_7**.
2. Select **Insert | Feature Operation | Chamfer**.
3. Click **Double Offset**.
4. Pick the edges to chamfer and click **OK**.
5. Enter values in the **First Offset** and **Second Offset** text box.
6. Click **OK**.

Exercise: Creating an offset angle chamfer

1. Continue with **mod_7**.
2. Select **Insert | Feature Operation | Chamfer**.
3. Click **Offset Angle**.
4. Pick the edges to chamfer and click **OK**.
5. Enter values in the **Offset** and **Angle** text boxes.
6. Click **OK**.

Offset Faces:

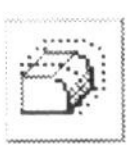 **Insert | Feature Operation | Offset Face**

Offset Value is a delta value relative to the current location of the face.

Offset Faces offsets individual faces.

Offset Features offsets every face that belongs to a selected feature.

Exercise: Creating offset faces

1. Open **mod_13**.
2. Select **Insert | Feature Operation | Offset Face**.
3. Enter a value in the **Offset Value** text box.
4. Click **OK**.
5. Click **Offset Faces** or **Offset Features**.
6. If you click **Offset Faces**, the **Offset Face** dialog displays prompting you to pick faces. Pick the faces to offset in the Graphics window and click **OK**.
7. If you click **Offset Features**, **Offset** dialog displays listing the available features in the model to offset. Pick the feature to offset from the list or in the Graphics window and click **OK**.

Note: You can enter a positive or negative value in the **Offset Value** text box. A positive value applies the offset in the direction of the surface normal (usually away from the solid) for each selected face or feature.

Hollow:

Hollow allows you to create a shelled solid body from a monolithic solid. The esulting body can have walls of different thickness. In addition, you can select faces as *pierce faces* and remove them from the hollowed body.

Three different Hollow types are available: **Face**, **Region** and **Body**. There are several options common to all hollow types.

A **Face** hollow creates a hollowed body from a single solid. You can define faces as pierce faces and remove them from the hollowed solid (to create openings and holes) and you can define hollowed faces with different thickness values.

A **Region** hollow is similar to a **Face** hollow, except you can define a group of pierce faces as a region, using seed and boundary faces, rather than selecting individual pierce faces. This is useful when there are a large number of adjacent faces to pierce.

A **Body** hollow differs from a **Face** hollow in that no pierce faces are defined. Use this to create a fully enclosed hollow body (like a beach ball). Solids created with a **Body** hollow can also have different wall thickness values.

Filter is used to limit the type of geometry available for selection by choosing **Face**, **Adjacent to Face** or **Body**.

Default Thickness is used to set the offset value for any faces in the solid that are not selected in **Offset Face**.

Alternate Thickness faces have their own thickness. Select the face from the list, and then enter a value in this text box.

Note: An entering negative value for a hollow feature adds material to the outside of the solid. Positive values remove material from the inside.

Creating a Face Hollow:

Insert | Feature Operation | Hollow

Exercise: Creating a face hollow

1. Open **mod_14**.
2. Select **Insert | Feature Operation | Hollow**. The **Hollow** dialog displays.
3. Click **Face**.
4. Click **Pierced Face**.
5. Pick the face(s) on the solid to pierce.
6. If necessary click **Offset Face** to define any faces with an alternate thickness value.
7. Pick a face on the solid that needs a different thickness value than the rest of the body.
8. Enter the value for this face in the **Alternate Thickness** text box.
9. Repeat steps 3 and 4 for each face needing an alternate thickness.
10. Enter a default hollow value in the **Default Thickness** text box.
11. Click **OK** or **Apply** to create the feature.

Creating a Region Hollow:

Insert | Feature Operation | Hollow

Exercise: Creating a region hollow

1. Open **mod_15**.
2. Select **Insert | Feature Operation | Hollow**.
3. Click **Region**.
4. Click **Seed Face**.
5. Pick a face from the group of faces to pierce.
6. Click **Boundary Face**.
7. Pick the faces that define the outer edge of the group of faces to pierce.

8. If necessary, click **Offset Face** to define any faces with an alternate thickness value.
9. In the Graphics window, pick a face on the solid that needs a different thickness value than the rest of the body.
10. Enter the value for these faces in the **Alternate Thickness** text box.
11. Repeat steps 9 and 10 for each face needing an alternate thickness.
12. Enter a default hollow value in the **Default Thickness** text box.
13. Click **OK** or **Apply** to create the feature.

Notes: I created a hollow using the Region option. Can I change it to a Body or Face hollow? Yes, select **Edit | Feature | Parameters** and choose the type you want.

Boundary faces cannot be tangent to any adjacent faces in the pierceregion.

Creating a Body Hollow:

Insert | Feature Operation | Hollow

Exercise: Creating a body hollow

1. Open **mod_16**.
2. Select **Insert | Feature Operation | Hollow**.
3. Click **Body.**
4. Pick the solid to hollow.
5. If necessary, click **Offset Face** to define any faces with an alternate thickness value.
6. Pick a face on the solid that needs a different thickness value than the rest of the body.
7. Enter the value for this face in the **Alternate Thickness** text box.
8. Repeat steps 4, 5 and 6 for each face that needs an alternate thickness
9. Enter a default hollow value in the **Default Thickness** text box.
10. Click **OK** or **Apply** to create the feature.

Taper:

Use **Taper** to add draft to faces on a solid body, relative to a specified vector. This is especially important on parts that you might manufacture via casting or injection molding.

There are four taper types: **Faces**, **From Edges**, **Tangent to Faces**, and **Split-line Taper**. In addition, you can create variable angle tapers using the **From Edges** method.

Several options are common to all taper types:

From Faces:

Creates a taper or draft on part faces, using a reference point to define the starting location of the angle. This type is generally best used on parts that have mostly straight edges and square faces.

From Edges:

This allows you to pick edges that are not in a plane normal to the taper direction vector. Unigraphics applies the taper to the faces sharing the edge, with the angle starting along the edge.

Variable Angle Taper:

A Variable Angle taper is a type of **From Edges** taper. You can specify different taper angles at points along the taper edges and Unigraphics interpolates the taper in between the points. This operation is similar to creating a variable radius edge blend.

Tangent to Faces:

Creates a taper tangent to cylindrical faces. This is useful in mold design because it allows for draft where undercuts may have occurred. You must specify a draw direction and taper angle but it is not required to specify a reference point.

Split - Line Taper:

Creates a taper on a subdivided face such that the tapered face is moved to lieon the reference point. This produces a step in the part at the split line from the tapered face over to the taper plane. This is very useful in plastic part design, where the mold split line must accommodate complex geometry. Along with the face edges, you must specify a draw direction, taper angle and a reference point.

Faces to Taper:

This step allows you to pick individual faces to taper and is used with **From Faces** and **Tangent to Faces** taper types. It is possible to select all the faces on a feature or body by setting the **Filter** to **Feature** or **Body**.

Reference Edges:

 This step allows you to pick the taper starting edges. The **Draw Direction** setting determines which of the adjacent faces are tapered. Use this selection step inconjunction with **From Edges** and **Split-line** tapers. Setting the **Filter** to **Face** selects all the edges on a given face.

Draw Direction:

 This step sets the vector that controls the taper direction. The taper **Angle** is defined relative to this vector.

Reference Point:

 This step allows you to pick the reference point or starting point for the taper. The "Reference Point" is a point on the taper plane through which the taper direction vector passes. The taper starts at this point and is applied in a direction based on the **Draw Direction** and **Angle** settings.
The **Reference Point** step is used on **From Faces** and **Split-line** tapers.

Angle sets the angle of the taper relative to the **Draw Direction**. A positive value creates an inward taper (toward the vector end point) and a negative value creates an outward taper (away from the vector end).

Taper All Instances will apply a taper to a form feature that belongs to an instance and have it automatically apply to all other members of the instance.

Face Taper:

 **Insert | Feature Operation | Taper**

Exercise: Creating a face(s) taper

1. Open **mod_17**.
2. Select **Insert | Feature Operation | Taper**.
3.  Click **Faces**.
4.  Click **Faces to Taper**.
5. Pick the faces to taper in the Graphics window.

6. Click **Draw Direction.**
7. Select a **Vector Method** from the list and then define the **Draw Direction**.
8. Click **Reference Point.**
9. Select a **Point Method** from the list and then define the **Reference Point**.
10. Enter a value in the **Angle** text box.
11. Click **OK** or **Apply** to create the taper.

Tapering from an Edge:

Insert | Feature Operation | Taper

Exercise: Creating a taper from edges

1. Reopen **mod_17**.
2. Select **Insert | Feature Operation | Taper** to display the **Taper** dialog.
3. Click **From Edges**.
4. Click **Reference Edges**.
5. Pick the edges from which to apply the taper.
6. Click **Draw Direction**. The **Taper** dialog changes allowing you to select a vector method.
7. Select a **Vector Method** from the list and then define the **Draw Direction**.
8. Enter a value in the **Angle** text box.
9. Click **OK** or **Apply** to create the taper.

Exercise: Creating a variable angle taper

1. Open **mod_18**.
2. Select **Insert | Feature Operation | Taper** to display the **Taper** dialog.
3. Click **From Edges**.
4. Click **Reference Edges**.
5. Pick the edges from which to apply the taper.
6. Click **Draw Direction**. The Taper dialog changes allowing you to select a vector method.
7. Select a **Vector Method** from the list and then define the **Draw Direction**.
8. Enter a value in the **Angle** text box.

9. Click **Variable Angle Point**. The dialog changes to show a list of variable angle points.
10. Select a **Point Method** from the list and then pick a **Variable Angle Point**.
11. Enter a value for this point in the **Variable Angle** text box.
12. Continue picking points and inputting values as needed.
13. Click **OK** or **Apply** to create the taper.

Notes: Can I make an existing taper feature into a variable angle taper? Yes, provided the taper was created as a From Edges taper. Select **Edit | Feature | Parameters** and add Variable Angle Points.

Variable Angle Points do not have to lie directly on reference edges. Unigraphics projects the points you select onto the reference edges.

Tangent to Face Taper:

Insert | Feature Operation | Taper

Exercise: Creating a taper tangent to a face(s)

1. Open **mod_19**.
2. Select **Insert | Feature Operation | Tape**.
3. Click **Tangent to Faces**.
4. Click **Faces to Taper**.
5. In the Graphics window pick the faces to taper.
6. Click **Draw Direction** The **Taper** dialog changes allowing you to select a vector method.
7. Select a **Vector Method** from the list and then define the **Draw Direction**.
8. Enter a value in the **Angle** text box.
9. Click **OK** or **Apply** to create the taper.

Notes: Can you add more faces to a taper already created?
Yes, edit the feature and add or remove the desired faces.

Use an edge or a datum axis to define the draw direction so it stays associative to the model.

Instance Arrays:

Rectangular:

Insert | Feature Operation | Instance

General creates an array from existing features and validates all geometry. When creating a general array the instance can to cross an edge of a face as well as move from one face to another.

Simple does not perform as much data validation or optimization so the array is built faster. Use this only if you are sure the instanced features will not interfere with edges from the target solid or other features.

Number Along XC controls the number of array columns, which Unigraphics creates parallel to the YC-axis of the WCS. The total number instanced includes the original feature.

XC Offset defines the spacing of the columns along the XC-axis.

Number Along YC controls the number of array rows, which Unigraphics creates parallel to the XC-axis of the WCS. The total number instance includes the original feature.

YC Offset defines the spacing of the rows along the YC-axis.

Exercise: Creating a rectangular array

1. Open **mod_20**.
2. Check the orientation of the WCS. If necessary use **WCS | Rotate WCS** to change the orientation. The XC-YC plane should be parallel to the desired array.
3. Select **Insert | Feature Operation | Instance**.
4. Click **Rectangular Array**. The **Instance** dialog displays with a list of all features in the model that you can instance.
5. Select the features to instance and click **OK**.
6. Select an array method and enter the array parameters.
7. Click **OK**. The instances display temporarily in the Graphics window and a dialog displays allowing you to create or reject them.
8. Click **Yes** to create the instance or **No** to return to the **Enter Parameters** dialog.

Notes: Can I instance blends or tapers? Not directly, but if you blend a form feature that has been instanced, you can add that blend to all the instances in the array by activating Blend All Instances. There are equivalent options on the Chamfer and Taper dialogs.
It is possible to leave out individual members of a rectangular array by suppressing the instances to skip; however, you cannot delete the instances.

Circular:

Insert | Feature Operation | Instance

General creates an array from existing features and validates all geometry. When creating a general array the instance can cross an edge of a face as well as move from one face to another.

Identical creates an exact copy of the feature but does not perform a validation check on the Boolean operation. This does not allow interferences with other edges or features.

Point Direction displays the **Vector Constructor** dialog to define the rotation axis.

Datum Axis displays a list of available datum axes in the model. Selecting a datum axis creates an associative axis of revolution for the circular array. This method is better than **Point Direction** for creating parametric, associative models.

Exercise: Creating a circular array

1. Open **mod_21**.
2. Select **Insert | Feature Operation | Instance.**
3. Click **Circular Array**. The **Instance** dialog displays with a list of all features in the model that you can instance.
4. Select the features to instance and click **OK**.
5. Select an array method and enter the array parameters.
6. Click **OK**.
7. Click **Point Direction** or **Datum Axis** and define the rotation axis for the array.
8. The instances display temporarily in the Graphics window and a dialog displays allowing you to create or reject them.
9. Click **Yes** to create the instance or **No** to return to the **Enter Parameters** dialog.

Note: Use **Datum Axis** when possible because it maintains associativity between the features and the array's axis of rotation.

Mirror Body:

 **Insert | Feature Operation | Instance**

Exercise: Creating a mirrored body

1. Open mod_**22**.
2. Select **Insert** | **Feature Operation** | **Instance**.
3. Click **Mirror Body**.
4. Pick the body or bodies to mirror and then click **OK**.
5. Pick a datum plane to mirror the bodies about.
6. Click **OK**.

Note: **Mirror Body** can be used as the last feature in a part to create one-half or one-quarter of a symmetrical part. This can improve the performance of large models.

Mirror Features:

 **Insert | Feature Operation | Instance**

Exercise: Creating a mirrored feature

1. Open **mod_23**.
2. Select **Insert** | **Feature Operation** | **Instance**.
3. Click **Mirror Feature**.
4. 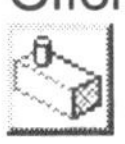 Click **Feature to Mirror**.
5. Select the features from the **Features in Part** list or pick them in the Graphics window.
6. Click **Add** to add features to the **Features in Mirror** list.
7. Click **Remove** to remove unwanted features from the **Feature in Mirror** list.
8.  Click **Mirror Plane**.
9. Pick the datum plane through which to mirror the features.
10. Click **OK** or **Apply** to create the **Mirror Set**.

Note: If possible, use a relative datum to mirror features rather than a fixed datum.

Editing Instances:

Instance features contain several options that you can only access by editing the feature after you create it. These include clocking instance arrays, suppressing individual instances, and changing the timestamp for mirrored bodies.

You can change circular or rectangular arrays by changing the array parameters or by clocking individual instanced features within the array. Select **Edit | Feature | Parameters** on the first instance in an array (INSTANCE [0]) to access both the feature dialog of the original instanced feature and the array dialog. Changing parameters in the feature dialog of any instance changes the parameters of all the instanced features.

Editing Array Positioning:

You can edit the positioning dimensions that place the initial feature used to create an array by selecting **Edit | Feature | Positioning** for the array feature, not the initial instance.

Editing Mirrored Bodies:

Use **At Timestamp** to change the point at which a body is mirrored. Selecting **Edit | Feature | Parameters** and activating **At Timestamp** allows the mirror feature to be reordered before other features in the original body. This can be useful if the mirrored body needs to be different from the original. If **At Timestamp** is inactive, Unigraphics automatically reorders the mirror feature to be the last feature in the model. If you add any features to the original body, the mirrored body includes these features.

At Timestamp is active by default when you create a mirror feature and you can only change it by editing the feature.

Editing Mirrored Features:

Features can be added to or removed from a mirror set by selecting **Edit | Feature | Parameters**.

Trim Body:

 **Insert | Feature Operation | Trim**

Define Plane defines a trimming plane using the **Plane** dialog.

Define Cylinder defines a cylinder as the trimming object. The two methods for this are **Diameter** and **Arc/Circle**.

Defines Sphere refines a sphere as the trimming object. The two methods for this are **Center, Diameter** and **Great Circle**.

Exercise: Creating a trimmed body

1. Open **mod_24**.
2. Select **Insert** | **Feature Operation** | **Trim**. The **Trim Body** dialog displays prompting you to select target bodies.
3. Pick the target body or bodies to trim and click **OK**. The **Trim Body** dialog displays prompting you to select a trimming object.
4. Pick the trimming object. If picking solid faces, click **OK** after you pick all the faces you need. If picking a datum plane or sheet body, the following dialog displays after you pick one object.
5. An arrow displays in the Graphics window showing the side of the body to be trim. Click **Accept Default Direction** or **Reverse Default Direction** to create the trim feature.
6. Click **Cancel**.

Notes: How often will I use a primitive (cylinder, sphere, cone or torus) to trim a body? Probably never, using sheet bodies and datum planes is much more common.

When trimming a body using a face, the face being specified must be large enough to pierce through the entire target body. If it is not large enough, a **Non-Manifold Solid**

Edit Trim Body:

 **Edit** | **Feature** | **Parameters**

Exercise: Editing a trim

1. Continue with **mod_24**.
2. Select **Edit** | **Feature** | **Parameters**.
3. Pick the desired **TRIM_SHEET** feature and click **OK**.
4. If you create a TRIM_SHEET feature by defining a plane, cylinder, sphere, cone or torus, the **Edit Trimmed Body** dialog displays allowing you to change the parameters of the trimming object.
5. If you create a TRIM_SHEET feature by picking an existing body, a dialog displays prompting you to edit the trimmed body. Click **Reverse Normal** to change the direction of the trim.
6. Click **Replace Surface** to replace the trimming surface. The **Trim Body** dialog displays, allowing you to pick a new surface.
7. Click **OK** to return to the **Edit Parameters** dialog.

8. Click **OK** or **Apply**.

Notes: Use **Delete Feature** to delete the TRIM_SHEET feature and restore the original solid shape.

Although a portion of the body has been trimmed away, Unigraphics remembers the topology of the entire solid body. Therefore, if you change the surface to a new plane or sheet body, provided the new plane/sheet can trim the original solid, the change updates.

Threads:

 **Insert | Feature Operation | Thread**

Thread Type defines the type of thread feature to create.

> **Symbolic Threads** represent a thread by adding a dashed arc to the hole or shaft. This creates a fast feature while still providing sufficient representation of the threads. Symbolic threads can use an external thread table file. This thread file contains all the common thread sizes and you can modify them to your specific standards. The thread standards available are: ANSI/Simplified, ANSI/Schematic, ANSI/Detailed, ISO/Simplified and ISO/Detailed.
>
> Symbolic threads are generally used to create threads for use in Drafting. After the symbolic thread is created you can display it in a drawing three ways: Detailed, Simplified and Schematic.
>
> **Detailed Threads** are complete, solid, helical cut threads, just as they would be on an actual threaded part. This creates a thread that looks more realistic but is significantly more complex (hence, slower) than a symbolic thread.
>
> **Major Diameter** is the largest diameter of the thread. This text box is available when you select an outside cylindrical face and an external thread is being created. Enter a minor diameter and the cylindrical face diameter is used as the major diameter.
>
> **Minor Diameter** is the smallest diameter of the thread. This text box is available when you select an inside cylindrical face and an internal thread is being created. Enter a major diameter and the cylindrical face diameter is used as the minor diameter.
>
> **Pitch** is the distance between the peaks of two threads, measured parallel to the axis of the cylindrical face.
>
> **Angle** is the included angle between the sides of the thread.

Callout is the name of the table entry from the table file. Use this to specify the thread form and size in Drafting.

Method specifies the thread type: **Rolled**, **Cut**, **Ground** or **Milled**.

Form offers different thread standards to choose from, such as Unified, American National, Metric, Acme, and others.

Full Thread creates the thread along the full length of the cylindrical face. Make this inactive to enter the thread length in the **Length** text box.

Length measures the thread length from the end of the selected cylindrical face closest to where you selected the face.

Manual Input disables the thread table and allows thread values to be entered manually.

Choose From Table allows you to select an entry from the thread table. If a size thread is selected from the table, the feature to which the thread is being applied changes to the size specified in the table. For example, if you pick a 1.0 diameter cylinder and apply a symbolic thread to it, you can select **Choose from Table** and select a 3/4-10 thread. The size of the cylinder changes and becomes linked to the thread feature.

Include Instances adds the thread to each instance if threads are added to a feature that is part of an instanced array.

Rotation allows you to specify a **Right Hand** or **Left Hand** thread. Select **Right Hand** to create the thread in a clockwise direction and select **Left Hand** to create the thread in a counterclockwise direction.

Select Start is used this to pick a solid planar face or datum plane as the starting location of the thread.

Exercise: Creating a symbolic thread

1. Open **mod_25**.
2. Select **Insert | Feature Operation | Thread**.
3. Select **Symbolic** as the **Thread Type**.
4. Pick the cylindrical face to thread. The previously grayed out parameters become available.
5. Enter the thread parameters or click **Choose from Table** to select a set of pre-defined parameters.
6. Click **OK** to create the symbolic thread feature.

Exercise: Creating a detailed thread

1. Open **mod_26**.
2. Select **Insert | Feature Operation | Thread**.
3. Select **Detailed** as the **Thread Type**.
4. Pick the cylindrical face to thread. The previously grayed out parameters become available.
5. Enter the thread parameters.
6. Click **OK** to create the detailed thread feature.

Notes: Can I edit threads after they have been created? Yes. Selecting **Edit | Feature | Parameters** brings up the Thread dialog, where you can access all of the parameters. However, you cannot make Detailed threads Symbolic or vice versa.

Pick a hole to create an internal thread and pick a shaft to create an external thread.

You can edit symbolic threads by selecting **Edit | Feature | Parameters**.
Because they use a helix, detailed threads take a long time to display. Use them only if displaying realistic looking threads is important.
Detailed threads may be shown on a drawing even if the thread feature is set to Symbolic.
Try to avoid using a threaded object in a Unite, Subtract or Intersect operation.

Editing Threads:

 **Edit | Feature | Parameters**

Exercise: Editing threads

1. Reopen **mod_25**.
2. Select **Edit | Feature | Parameters**.
3. Select the **SYMBOLIC_THREAD** or **THREADS** feature from the list, the Graphics window. Click **OK** to display the **Edit Parameters** dialog and edit the feature parameters.
4. Click **Feature Dialog** to display the **Edit Thread** dialog.
5. Enter the new thread parameters as needed and click **OK**.

Note: You cannot reattach a thread feature to a new face.

Solid Modeling:

Solid Modeling is the process of creating and modifying solid objects. Each solid object has topology defining the surface area and volume of an object. Topology is comprised of faces, edges, and vertices.

One of the advantages of solid modeling is the ability to analyze the model. Detailed mass properties (weight, volume, moment of inertia, and more) can be calculated for a solid body, as shown in the Information window.

Analysis can be as simple as mass properties calculations or as complicated as dynamic stress/strain or fluid flow calculations.

Some advantages of solid modeling include:

- Creation of a "virtual solid," geometrically identical to the actual part
- Ability to rapidly create and change models
- Powerful feature associativity and feature positioning
- Greater ease in modifying designs
- Substantially improved design visualization
- Mass properties information
- Easy to create rapid-prototyping output
- Evaluation of form, fit, and function
- Using an assembly to design new components in the context of an entire product model
- Checking interferences between parts

Design Intent:

In traditional solid modeling, steps used to build the solid body are not recorded. This is referred to as explicit solid modeling. The solid body is created without capturing a history of the individual features used to construct it. This makes the process of modifying the solid body much more complex and time consuming.

"**Design Intent**" is the ability to capture geometric conditions, dimensional relationships, and relative positions between features on solid models.

A CAD software application that can capture design intent is called a "parametric modeler." Unigraphics has powerful parametric modeling functions. The program also gives you a choice to add, or later remove, design intent. At the start of a design, you may not know what sort of design intent is necessary. Using expressions, positional dimensions, instances and feature editing, you can change or add design intent to the solid model.

Parametric Modeling:

Design intent is captured through parameters. Parameters are all of the necessary dimensions used to describe the various features in a solid model.

These dimensions are called Feature Parameters. The dimensions used to locate the feature are called Positioning Dimensions.

These parameters describe the feature in the solid model. When a change to the solid model is required, the parameters associated with a particular feature are modified.

Each parameter is assigned a name by Unigraphics. The names start at "**p0**" and increment by one.

Feature Based Modeling:

Unigraphics presents an approach to modeling called "feature based design." A feature is a tool used to create and shape a solid model. There are many different types of features. A feature based solid model is a collection of these features. The solid is much easier to describe when it is broken down into the individual features used to construct it.

Each of the features used in building a solid model is assigned a feature number, as shown in the Model Navigator feature list.
The first feature starts with "0" (i.e., BLOCK (0)). This feature is the base feature of the solid. Each time a new feature is added, the feature number increases by one.

The feature order describes the way the part was built. It gives the history of the solid body's construction. In the example, an extrusion was created first, followed by a blend, and then a rectangular slot.

Editing or Modifying Feature Based Models:

In order to demonstrate feature-based design modifications consider a moderately difficult design change. Imagine that most of the design in the model is created, taking into account blends and material thickness. Now there is a last-minute change in the design due to the results of a finite element analysis. It is determined the wall thickness needs to be increased to prevent failure. The size of the blends must also increase to make the design more aesthetically pleasing.

If the model were a curves and surfaces model, the changes could take many hours to implement. You would probably need to delete and recreate objects. With explicit solid modeling, these changes could also take hours because you would need recreate solids or manually edit faces. In feature-based design, the changes can be done in minutes by altering the parameters used to define the features.

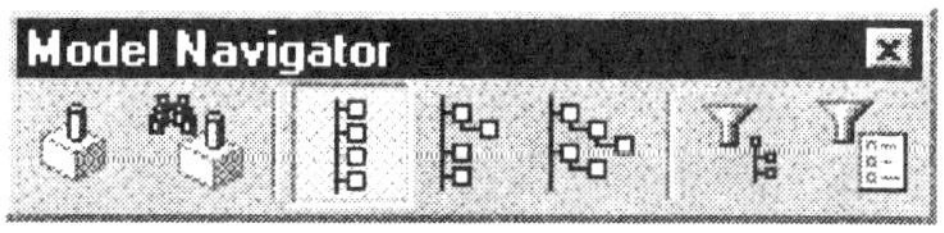

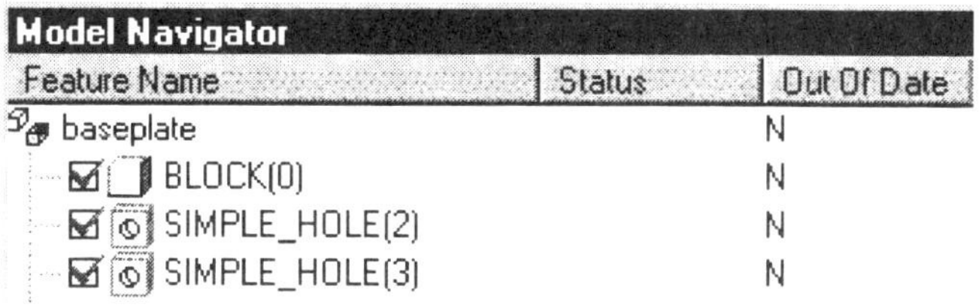

Model Navigator, (MNT) is another method of editing the features of a parametric model by selecting **View | Model Navigator**.

The Model Navigator provides a simple and convenient way to view the features in a model and quickly edit them.

The Model Navigator graphically displays the features in a model. The example shown displays in "Time Stamp Order," showing the order in which the features apply to the model. This is the same order you see by selecting **Edit | Feature | Parameters**.

To modify a feature, right-click over it in the Model Navigator, and select **Edit Parameters** from the menu.
This produces the same result as selecting **Edit | Feature | Parameters**, picking the feature and clicking **OK**.

The Model Navigator has other powerful features. Use it to quickly control the update status of features (suppression), change the order in which features apply to the model, and view dependency relationships among features. It can also provide detailed information about the model, its features, and their parameters.

Primitives Features are analytical shapes useful as basic building blocks in solid modeling. You can modify parameters to change their size and shape, but not their positioning. **Block, Cylinder, Cone** and **Sphere** are the four types of primitive features.

Primitives cannot be located with relative positioning constraints. You must place them using coordinate point locations. This limits the effectiveness of primitives when a solid already exists in the part file because a positioning relationship cannot be established.

Note: To build a fully parametric model, only one primitive should be used (as the first feature).

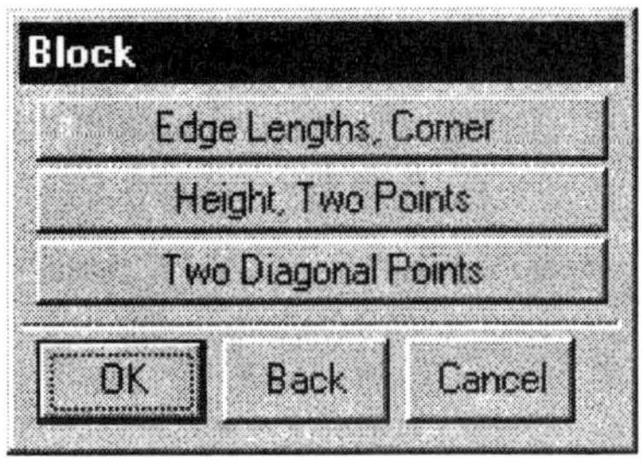

Creating a Block:

 **Insert | Form Feature | Block**

Edge Lengths, Corner creates a block using a height value and two diagonal points to determine the length and width with respect to the WCS.

The height follows the orientation of the +ZC axis. The **Point Constructor** is used to specify two points to determine the length (+XC) and width (+YC). Unigraphics projects the two points onto the XC-YC plane using a normal vector. The two points determine the diagonal of one side of the block. The origin of the block is at the first point specified.

Two Diagonal Points creates a block using two diagonal points to determine the length, width, and height with respect to the WCS.

The **Point Constructor** is used to specify two diagonal points to determine the length (+XC), width (+YC), and height (+ZC). The first point specified defines the origin of the block. The second point cannot have the same ZC value as the first point.

Exercise: Creating a block using the edge, length and corner method

1. Open **seed_in**.
2. Select **Insert** | **Form Feature** | **Block**.
3. Click **Edge Lengths, Corner**.
4. Enter edge lengths in the **X**, **Y**, and **Z** text boxes. Click **OK**. The **Point Constructor** dialog displays.
5. Define the Block corner location, and click **OK**.

Note: Only positive length values can be entered for the Block parameters (Edge Lengths and Height).

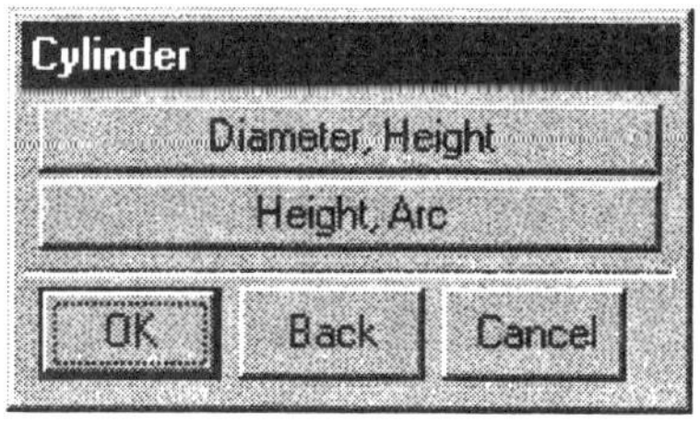

Creating a Cylinder:

 **Insert | Form Feature | Cylinder**

Diameter, Height creates a cylinder using a direction vector, a diameter value, a height relative to the direction vector, and a location. The height must be a positive value. The **Vector Constructor** dialog displays to define the centerline and direction of the cylinder.

The **Point Constructor** dialog is where you specify an origin location for the cylinder. The origin is the center of the circular face at the bottom of the cylinder.

Height, Arc creates a cylinder using an arc to determine the diameter, direction vector, and location of the cylinder. The height must be a positive value.
An arc defines the cylinder's location and diameter. The cylinder is not associated to the arc.

Exercise: Creating a cylinder using the diameter and height method

1. Open **seed_mm**.
2. Select **Insert | Form Feature | Cylinder**.
3. Click the **Diameter, Height** icon.
4. Define the direction of the cylinder axis, and click **OK** to display the **Cylinder** dialog (for parameters).
5. Enter the **Diameter** and **Height** of the cylinder, and click **OK** to display the **Point Constructor** dialog.
6. Define the cylinder's origin point, and click **OK**. If no other solids exist in the model, the cylinder is created. Otherwise, the **Boolean Operation** dialog displays.
7. Select the appropriate Boolean operation to create the cylinder.

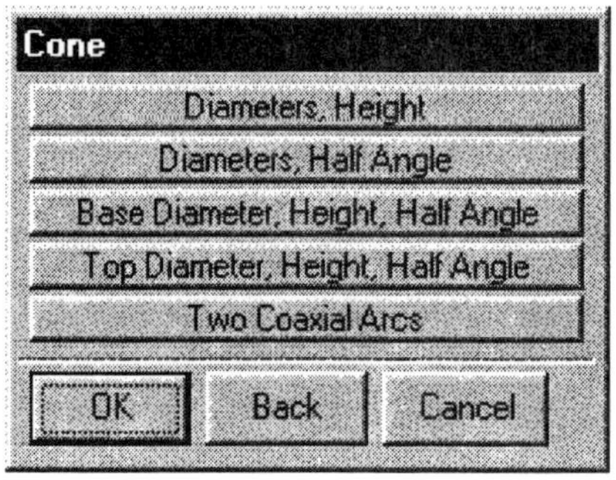

Creating a Cone:

 **Insert | Form Feature | Cone**

Diameters, Height creates a cone using a direction vector, base and top diameter values, a height relative to the direction vector, and a location. The height value must be positive. Use this when you do not know the half angle. The origin of the cone is the center of its bottom circular face.

Diameters, Half Angle creates a cone using a similar procedure as the **Diameters, Height** option. The difference is that a half angle value determines the height Use this method if the height is unknown or is not the driving parameter. The half angle must be greater than 0-degrees and less than 90-degrees.

Base Diameters, Height, Half Angle creates a cone using a similar procedure as the **Diameters, Height** option. The difference is that the height and half angle values determine the top diameter. Use this if the top diameter is unknown or is not the driving parameter.

The half angle must be between 0- and 90-degrees or between 0- and –90-degrees. If the half angle is negative, the top diameter is larger than the base diameter. The height must be a positive value.

Top Diameter, Height, Half Angle creates a cone using a similar procedure as the **Diameters, Height** option. The difference is that the height and half angle values determine the base diameter. Use this if the base diameter is unknown or is not the driving parameter. The half angle must be between 0- and 90 - degrees. The height must be a positive value.

Two Coaxial Arcs creates a cone defined by two existing arcs. The base arc defines the base diameter and the cone axis orientation. The top arc defines the top diameter. The cone axis is defined normal to the base arc. The two arcs determine the diameters of the cone base and top. The distance between the arc center of the top arc and the base arc determines the cone height.

If the selected arcs are not coaxial, Unigraphics intersects the cone's axis with a plane parallel to the base arc containing the center point of the top arc. The distance between the base arc center point and the intersection point defines the height of the cone. The top arc diameter determines the top diameter of the cone. Unigraphics projects the top arc center point normal to the cone axis to find an intersection. This intersection point determines the height of the cone. The diameter of the top arc then defines the top of the cone until it is parallel to the plane of the base arc.

Exercise: Creating a cone using the diameters and height method

1. Open **seed_in**.
2. Select **Insert | Form Feature | Cone** to display the **Cone** dialog (for creation method).
3. Click **Diameters, Height** to display the **Vector Constructor** dialog.
4. Define the axis of the cone and click **OK** to display the **Cone** dialog (for parameters).
5. Enter the desired values in the **Base Diameter**, **Top Diameter**, and **Height** text boxes.
6. Click **OK** to display the **Point Constructor** dialog.
7. Define the origin of the cone and click **OK**. If no other solids exist in the model, the cylinder is created. Otherwise, the **Boolean Operation** dialog displays.
8. Select the appropriate Boolean operation to create the cylinder.

Exercise: Creating a cone using the diameters and half angle method

1. Continue with **seed_in**.
2. Select **Insert | Form Feature | Cone**.
3. Click **Diameters, Half Angle**.
4. Define the axis of the cone and click **OK**.
5. Enter the desired values for the **Base Diameter**, **Top Diameter**, and **Half Angle**.
6. Click **OK** to display the **Point Constructor** dialog.
7. Define the origin of the cone and click **OK**. If no other solids exist in the model, the cylinder is created. Otherwise, the **Boolean Operation** dialog displays.
8. Select the appropriate Boolean operation to create the cone.

Note: A **Top Diameter** of **0** produces a sharp-pointed cone.

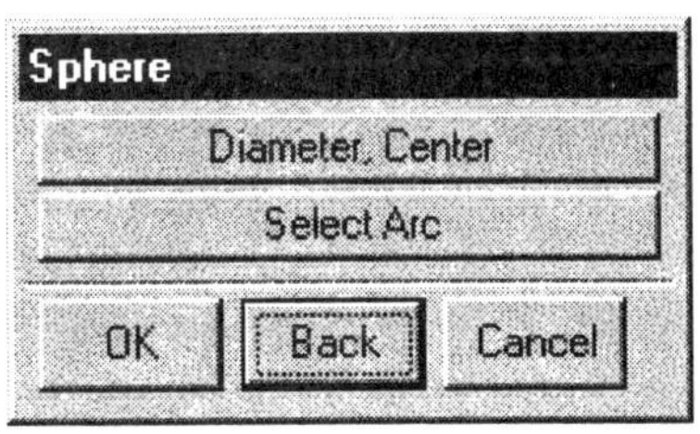

Creating a Sphere:

Insert | Form Feature | Sphere

Diameter, Center creates a sphere using a specified diameter. The **Point Constructor** displays to allow you to specify the origin location for the sphere.

Select Arc creates a sphere using the diameter of a selected arc. The center of the selected arc is the sphere origin. The sphere is not associated to the arc.

Exercise: Create a sphere using the diameter and center method

1. Open **seed_mm**.
2. Select **Insert | Form Feature | Sphere**.
3. Click **Diameter, Center**.
4. Enter a diameter and click **OK**.
5. Define the sphere center and click **OK**. If no other solids exist in the model, the cylinder is created. Otherwise, the **Boolean Operation** dialog displays.
6. Select the appropriate Boolean operation to create the cylinder.

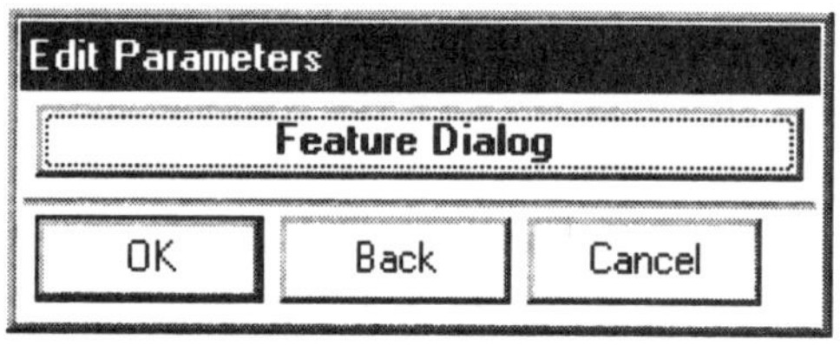

Editing and Positioning Primitive Features:

The parameters describing primitive features, such as diameters, heights, and angles, can be edited after the feature creation.

Selecting **Edit | Feature | Parameters** for a primitive feature (block, cylinder, cone, sphere) allows you to edit the sizing parameters used in the feature's creation. Click **Feature Dialog** to change these parameters. The same dialog used to create the feature displays.

The main disadvantage to using primitive features throughout a model is their inability to be parametrically repositioned. Having no positioning constraints, you cannot move them easily once placed. It is possible to reposition a primitive using **Edit | Feature | Move**, but you must have the WCS oriented correctly to use this option.

For this reason, primitives should generally only be used as the initial feature in a model. After that, it is better to use extruded and revolved features (or bosses, pads, pockets, and holes) to create primitive shapes. These parametric form features are just as easy to create as primitive features, but their parametric positioning capabilities make them infinitely more powerful.

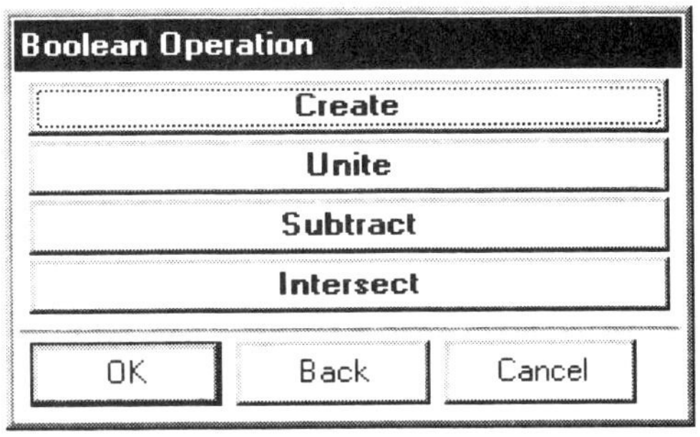

Boolean Operations:

Unite:

 **Insert | Feature Operation | Unite**

Exercise: Uniting solids

1. Open **mod_27**.
2. Select **Insert | Feature Operation | Unite**.
3. Pick the target solid from the Graphics window. The **Class Selection** dialog displays.
4. Pick the tool solid(s), and click **OK** to unite them.

Notes: Can I change the bodies used in the operation after it has been performed? Yes, use **Edit | Feature | Parameters** to change the tool and target body.

Sheet bodies cannot be united.

Subtract:

 **Insert | Feature Operation | Subtract**

Exercise: Subtracting solids

1. Open **mod_28**.
2. Select **Insert | Feature Operation | Subtract**.
3. Pick the target solid in the Graphics window, the **Class Selection** dialog displays.
4. Pick the tool solid(s) and click **OK** to subtract them.

Notes: Can I change the bodies used in the operation after it has been created? Yes, use **Edit | Feature | Parameters** to change the tool and target body.

The target body for **Subtract** can be a sheet body. A sheet body can also be the tool body, provided the target body is a sheet, and the two bodies have coincident faces. It is often a good idea to perform an **Extract** on the tool solid, and then subtract the extracted solid from the target solid. Otherwise, the tool solid cannot be edited once it has been subtracted.

Intersect:

Insert | Feature Operation | Intersect

Exercise: Intersecting solids

1. Open **mod_29**.
2. Select **Insert | Feature Operation | Intersect**.
3. Pick the target solid from the Graphics window, the **Class Selection** dialog displays.
4. Pick the tool solid(s) to intersect. Click **OK**.

Notes: Can I change the bodies used in the operation after it is created? Yes, use **Edit | Feature | Parameters** to change the tool and target body.

Intersect can use sheet bodies, as long as they have coincident faces. You can pick just sheet bodies as the tool or target. If you intersect a sheet body with a solid body, you must pick the sheet body as the target.

It is a good idea to create all the possible shapes before performing the Boolean operation and then use **Edit Parameters** to swap the tool and target bodies.

You can use sheet bodies and solid bodies for the new target body or tool body. The timestamp of the body used in the edit must be before the Boolean feature.

Non-Manifold Conditions:

When using the Subtract operation, the edges, or vertices of the new feature/body, may not touch the target solid such that any part of the resultant solid would have a zero thickness (e.g., the cylindrical face of the hole and the external planar face are coincident).

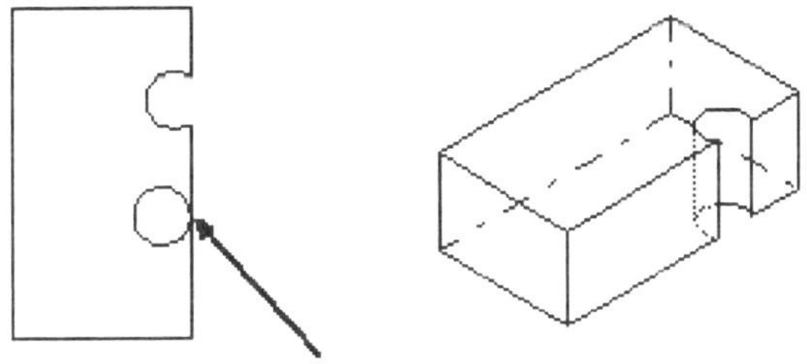

A zero thickness (non-manifold condition) is not allowed.

If a non-manifold condition occurs, the system will not perform the operation and you will receive a message as follows: Non-Manifold Solid.

Transformations:

Translate:

Edit | Transform

To a Point allows you to move or copy geometry from one point to another using the **Point Constructor** dialog. You must first define a Reference Point and then define a Destination Point. The Reference Point is the base point or the origin of the transformation. The Destination Point is the point where the Reference Point and all selected objects are moved or copied. The Reference and Destination Points do not have to be on or near the objects.

Delta allows you to input distances along the XC, YC and ZC axes of theWCS. You can consider the 0,0,0 coordinate to be the Reference Point and the Delta values entered to be the coordinates of the Destination Point.

Exercise: Translating an object to a point

1. Open **mod_30**.
2. Select **Edit | Transform**.
3. Pick the object(s) to transform, and click **OK**.
4. Click **Translate**.
5. Click **To a Point**.
6. Define the Reference Point and click **OK**.
7. Define the Destination Point and click **OK**.
8. Select any necessary transformation options.
9. Select **Move** or **Copy** to transform the object(s).
10. Click **Cancel** to finish, or click **OK** to translate the object(s) again.

Exercise: Translating an object a distance

1. Reopen **mod_30**.
2. Select **Edit | Transform**.
3. Pick the object(s) to transform, and click **OK**.
4. Click **Translate**.
5. Click **Delta**.
6. Enter the Delta values and click **OK**.
7. Select any necessary transformation options.
8. Select **Move** or **Copy** to transform the object(s).
9. Click **Cancel** to finish, or click **OK** to translate the object(s) again.

Rotate about a Point:

Edit | Transform

Angle allows you to type in a rotation angle. The object rotates in the XC-YC plane about an axis that is parallel to the +ZC axis and through the Reference Point specified. The positive rotation direction is counterclockwise.

Two Point Method requires you to select two points to determine the angle and direction of rotation. The geometry rotates *from* Point 1 *to* Point 2 about the reference point using the Z-axis of the WCS as the rotational axis.

Exercise: Rotating an object about a point

1. Open **mod_31**.
2. Select **Edit | Transform**.
3. Pick the object(s) to transform, and click.
4. Click **Rotate About a Point**.
5. Define the point of the center of rotation and click **OK**.
6. Enter the rotation angle and click.
7. Select any necessary transformation options.
8. Click **Move** or **Copy** to transform the object(s).
9. Click **Cancel** to finish, or click **OK** to translate the object(s) again.

Exercise: Rotating an object about a point (two point method)

1. Reopen **mod_31**.
2. Select **Edit | Transform**.
3. Pick the object(s) to transform, and click **OK**.
4. Click **Rotate About a Point**.
5. Define the point of the center of rotation and click **OK**.
6. Click **Two Point Method**.
7. Define the first point and click **OK**.

8. Define the second point and click **OK**.
9. Select any necessary transformation options.
10. Click **Move** or **Copy** to transform the object(s).
11. Click **Cancel** to finish, or click **OK** to translate the object(s) again.

Mirror through a Line:

Edit | Transform

Two points requires two points to define the line through which to mirror the selected objects.

Existing Line requires a linear curve or edge through which to mirror the selected objects.

Point and Vector requires an origin point and a linear direction. This defines the line through which to mirror the selected objects.

Exercise: Mirroring an object through a two points

1. Open **mod_32**.
2. Select **Edit** | **Transform**.
3. Pick the object(s) to transform, and click.
4. Click **Mirror Through a Line**.
5. Click **Two Points**.
6. Define the first point of the mirror line and click **OK**.
7. Define the second point of the mirror line and click **OK**.
8. Select any necessary transformation options.
9. Click **Move** or **Copy** to transform the object(s).
10. Click **Cancel** to finish, or click **OK** to translate the object(s) again.

Exercise: Mirroring an object through a line

1. Continue with **mod_32**.
2. Select **Edit** | **Transform.**
3. Pick the object(s) to transform, and click **OK**.
4. Click **Mirror Through a Line**.
5. Click **Existing Line**.
6. Pick the line through which to mirror.
7. Select any necessary transformation options.
8. Click **Move** or **Copy** to transform the object(s).
9. Click **Cancel** to finish, or click **OK** to translate the object(s) again.

Exercise: Mirroring an object through a point and vector

1. Continue with **mod_32**.

2. Select **Edit | Transform**.
3. Pick the object(s) to transform, and click.
4. Click **Mirror Through a Line**.
5. Click **Point and Vector**.
6. Define the start point of the line and click **OK**.
7. Define the direction of the line and click **OK**.
8. Select any necessary transformation options.
9. Click **Move** or **Copy** to transform the object(s).
10. Click **Cancel** to finish, or click **OK** to translate the object(s) again.

Note: You cannot mirror through a line that is normal to the work plane because there is an infinite number of temporary planes that Unigraphics could create. Use **Undo Last** to remove the effects of the transformation if the specified line was normal to the work plane.

Rotate about a Line:

Edit | Transform

Rotate About a Line rotates objects in 3D space around a linear object acting as the axis; the rotation angle is counterclockwise, as measured along the positive axis direction. This is unlike **Rotate About a Point** that rotates objects in the XC-YC plane.

Two Points requires two points to define a line to rotate the selected geometry around. The first point defines the vector origin and the second point defines the vector direction and orientation.

Exercise: Rotating an object about a line, two points

1. Open **mod_33**.
2. Select **Edit | Transform**.
3. Pick the object(s) to transform, and click **OK**.
4. Click **Rotate About a Line**.
5. Click **Two Points**.
6. Define the first point of the rotation line and click **OK**.
7. Define the second point of the mirror line and click **OK**.
8. Enter the rotation angle and click **OK**.
9. Select any necessary transformation options.
10. Click **Move** or **Copy** to transform the object(s).
11. Click **Cancel** to finish, or click **OK** to translate the object(s) again.

Existing Line requires a linear curve or edge to rotate the selected objects around. The point at which the linear segment is selected determines the vector origin and direction. The endpoint nearest the linear segment selection point is the vector origin and the opposite endpoint establishes the vector direction and orientation.

Exercise: Rotating an object about a line

1. Open **mod_33**.
2. Select **Edit | Transform**.
3. Pick the object(s) to transform, and click **OK**.
4. Click **Rotate About a Line**.
5. Click **Existing Line**.
6. Pick the line about which to rotate.
7. Enter the rotation angle and click.
8. Select any necessary transformation options.
9. Click **Move** or **Copy** to transform the object(s).
10. Click **Cancel** to finish, or click **OK** to translate the object(s) again.

Point and Vector requires an origin point and a linear direction. This defines the line that the selected objects rotate around.

Exercise: Rotating an object around a point and vector

1. Reopen **mod_33**.
2. Select **Edit | Transform**.
3. Pick the object(s) to transform, and click **OK**.
4. Click **Rotate About a Line**.
5. Click **Point and Vector**.
6. Define the start point of the line and click **OK**.
7. Define the direction of the line and click **OK**.
8. Select any necessary transformation options.
9. Click **Move** or **Copy** to transform the object(s).
10. Click **Cancel** to finish, or click **OK** to translate the object(s) again.

Note: This operation creates an unparameterized feature if used on a parametric body.

Reposition:

Edit | Transform

Exercise: Transforming an object using reposition

1. Open **mod_34**.
2. Select **Edit | Transform**.
3. Pick the object(s) to transform, and click **OK**.
4. Click **Reposition**.
5. Define the Reference CSYS.
6. Define the Destination CSYS.
7. Select any necessary transformation options.
8. Click **Move** or **Copy** to transform the object(s).

9. Click **Cancel** to finish, or click **OK** to translate the object(s) again.

Rotate Between Two Axis:

Edit | Transform

Minor/Major Angle is created through a Reference Point and normal to a plane defined by the reference axis and the destination axis. The angle between these two axes is the positive **Minor Angle**. If you want to rotate the objects in the reverse direction, select **Major Angle** and input the desired angle.

Exercise: transforming an object with rotate between two axes

1. Open **mod_35**.
2. Select **Edit | Transform**.
3. Pick the object(s) to transform, and click **OK**.
4. Click **Rotate Between Two Axes**.
5. Define the Reference Point.
6. Define the Reference Vector.
7. Define the Destination Vector.
8. Click **Major Angle** and enter an angle value and click **OK**.
9. Select any necessary transformation options.
10. Click **Move** or **Copy** to transform the object(s).
11. Click **Cancel** to finish, or click **OK** to translate the object(s) again.

Summary:

In this chapter you learned how to:

- Recognize the Advantages of Using Solid Models.
- Capture Design Intent in a Model.
- Use Parameters to Define a Feature's Size and Position.
- Recognize the Benefits of the Feature-based Approach to Building Solid Models.
- Edit Basic Feature Parameters, and the Expressions Controlling Them.
- Use the Model Navigator to Edit Features.
- Use Section Strings with Swept Features.
- Perform Boolean Operations with Swept Features.
- Extrude Section Strings Using Various Methods
- Revolve Section Strings using Various Methods.
- Sweep a Section String Along a Guide.
- Create a Tube or Cable Feature.
- Edit Swept Features.
- Use Primitive Bodies Properly.
- Create Primitives in the Shape of Blocks, Cylinders, Cones, and Spheres.
- Edit Primitive Bodies.

- Use Boolean Operations.
- Unite Multiple Bodies.
- Subtract Bodies From a Target Body.
- Intersect Multiple Bodies.
- Edit Boolean Operations.
- Transform Objects.
- Move Objects Via Translation.
- Rotate Objects about a Point.
- Mirror Objects through a Line.
- Create a Rectangular Array of Objects.
- Create a Circular Array of Objects.
- Rotate Objects about a Line.
- Mirror Objects through a Plane.
- Reposition Objects Using Coordinate Systems.
- Determine When to Use Form Features.
- Use Relative Positioning to Locate Form Features on a Model.
- Create a Boss.
- Create Various Holes.
- Create Different Shaped Pads.
- Create Different Shaped Pockets.
- Create Various Slots.
- Create Different Grooves.
- Edit Form Features After You Create Them.
- Create Constant Radius Edge Blends.
- Create Variable Radius Edge Blends.
- Create Cliff Edge Blends.
- Handle Overflow Conditions in Edge Blends.
- Edit Edge Blend Parameters.
- Chamfer edges of a model.
- Parametrically Offset Solid Faces to Thicken Walls.
- Hollow Solid Bodies.
- Add Draft to a Model by Tapering Faces.
- Create Arrays to Define a Pattern of Features.
- Mirror Solid Bodies to Create a Right or Left-hand Copy of a Model.
- Mirror Symmetrical Features on a Model.
- Trim Bodies by Using Sheets or Planes.
- Create Symbolic and Detailed Threads

Chapter 5

Assembly Modeling

After completing this chapter, you will be able to:

- Understand and Apply the Assembly Structure Concept.
- Distinguish between the Work Part and the Displayed Part.
- Change the Work Part.
- Change the Displayed Part.
- View the Assembly Structure and Use the Assembly Navigator to Make Selections.
- Find Components and Perform Operations on Them.
- Create and Edit an Assembly Using the Bottom-Up Approach.
- Create and Manage Reference Sets.
- Display Only Necessary Geometry in Your Assembly by Replacing Reference Sets.
- Create a Top-Down Assembly.
- Use the Top-Down Approach to Design in Context of the Assembly.
- Create Sketches in the Assembly.
- Recognize the Importance of the Work Part.
- Remove Components.
- Transform, Rotate and Reposition Components.
- Understand the Mating Components Concept in an Assembly and Its Advantages.
- Create Mating Conditions and Constraints.
- Recognize Common Problems Encountered with Mating Conditions and Solve Them.
- Create an Explosion, Either Automatically or Manually.
- Show or Hide an Explosion from a View.
- Distinguish Between Partially and Fully Loaded Parts.
- Use Load Options to Manage Large Assemblies.

Assemblies:

The Assemblies application allows you to create a model containing multiple parts. You can create and maintain libraries of components that may be common to many parts, develop part designs within the context of the assembly, create assembly drawings and parts lists, and monitor changes that occur within the assembly

The creation of assembly files is recommended for a number of reasons. By using assemblies, your overall file size is smaller because the assembly file creates a link to the various parts used in the assembly, rather than making multiple copies of each part. When a part is used multiple times in an assembly (such as bolts or pins), the part file is referenced once, and multiple "occurrences" of that part are displayed in the assembly.

Another advantage to assemblies is that individual part files are "master" files, and changes to these files update assembly and drafting files, instead of having to make the same change multiple times. Using methods such as Interpart Expressions can extend that capability by linking common parameters between part files. In this way, you can make an edit to a bolt diameter, and have the thru hole diameter in other parts change simultaneously. Unigraphics allows multiple part files to be open at the same time, so you can see changes to all the parts immediately.

Assemblies also allow you to create relationships between the different parts through the use of "mating conditions." Mating conditions allow you to establish smart relationships between adjacent parts. For example, if a part is changed in a way that makes a hole position move, the bolt mated to that hole also moves.

You create assemblies by adding references to master parts into a new part file. These references are referred to as "components." The assembly is modified by editing the master part files, by adding or removing components, or by replacing components with alternate versions of the master part. There are several methods available to control the loading and display of components within the assembly.

There are six terms you need to understand to work with assemblies: Top-Down, Bottom-Up, Work Part, Displayed Part, Loaded Part and Design in Context.

Top Down assembly modeling allows you to create and/or edit component geometry (piece parts) at the assembly level. Since creation and editing is performed in the assembly, you can take advantage of each existing component's geometry and location in the design of new component parts. All creation and/or editing done at the component level is automatically reflected in the component's part file.

Bottom Up implies that all of the individual part files (piece parts) making up the assembly already exist. You simply add these existing parts to create the assembly. Bottom-Up modeling is analogous to building a toy model, the pieces are already created, and you only need to assemble them.

When you edit an individual component part file outside of the assembly structure, the changes are automatically reflected in each assembly structure that contains the component.

Work Part is the part in which you create or edit geometry. When you are working in a component part, the Work Part is the same as the Displayed Part. When working in an assembly, you can make one of the components your Work Part. All other components become temporarily unavailable. The Work Part name displays in the title bar of the Graphics window.

The Work Part is the part in which you are actually working. Use **Assemblies | Change Work Part**, or double-click the part in the Assembly Navigator to make a part your Work Part. All new geometry is created in the Work Part. When you save a file, the Work Part and all components of the Work Part are saved. The Work Part is also the part file to which components are added, removed, repositioned, renamed, and substituted.

Displayed Part is the part currently displayed in the Graphics window. The Displayed Part name appears in the header of the Graphics window.

Loaded Part is any open part. When an assembly file is opened, all of its components also open when the **Load Components** option is set to **All Components** using **File | Options | Load Options**. You can find out which parts loaded by reviewing the contents of the list in the **Change Displayed Part** dialog by using **Window | More Parts**.

Designing in Context:

You have the ability to create and edit components while in the assembly. This is known as designing the part in the context of the assembly. This approach causes the change to occur simultaneously at both the assembly and the component level. In order to modify a component in the assembly file, you must first make the component your Work Part.

Note: One very important rule in understanding assemblies is you must always know which component is the Work Part.

The Assembly Structure:

Understanding the following concepts makes it much easier for you to build and maintain the most optimal assembly structures.

The assembly file is nothing more than a part file that contains component geometry. It is created using **File | New** or **Assemblies | Edit Structure | Create**. Think of the assembly file as a collection of components and other sub-assemblies with their own components. Assemblies can have many levels, where each level contains components of some upper level assembly.

Components are the individual part files making up an assembly. Adding a part file to an assembly structure creates a component. Components can have unique names, reference sets, colors, layers, and attributes. Use them to identify different piece parts in an assembly.

For example, a single plate can be designed for use as both an upper and lower plate in an assembly (i.e., a coupling). The plate is added to the assembly twice, one plate can have the component name "top_cover" and the other "bottom_cover". Although they have different component names, there is only one single cover part file defining them. Changes to that part file from within the assembly, or outside of the assembly, causes both the top and bottom covers to update.

Component geometry is not copied into the assembly file when creating or adding components. When the assembly structure is defined and components created or added into the assembly, instances are created. An instance is not a duplication of geometry, rather it is an object linked to the original object. Since the original geometry exists in the component part file any changes to the original geometry, made inside the assembly or outside of the assembly, are automatically reflected in all of the instances. This is what enables top down and bottom up associativity to maintain automatically. Therefore it is possible, even likely, an assembly file only contains instances, assembly drawings and manufacturing information.

When an assembly part is loaded, you can use these instances to locate and load all of the component parts into memory. The component instances are displayed in the assembly part file. They allow the assembly file to be quite small in size as well as maintain automatic associativity between the assembly structure and the individual component parts.

If a single component is used multiple times within the same assembly, there are multiple occurrences of the instance. Again, a change made to any occurrence is automatically made in all of the instance occurrences.

For example, in the design of a machine, many bolts of the same size are used. If a machine design uses 150 ½" X 2" bolts, you create a single component part file to define the bolt. When you add a bolt into the machine assembly structure, you establish the instance. Unigraphics adds the rest of the bolts as instance occurrences of the bolt component part. Since geometry is not copied, the size of the assembly is much smaller and changing the single bolt component part file updates all 150 bolts. It doesn't matter if you select the first or 150th occurrence; the update is automatic to all 150 bolts even if you change the bolt part file outside of the assembly!

Work Part vs. Displayed Part:

When you initially load an assembly, the Work Part and the Displayed Part is the same. It is important to understand all additions and changes made are always made in the Work Part. You can change the Work Part to another subassembly part or an individual component. The Displayed Part can be the whole assembly, a subassembly, or an individual component.

Using the previous example, change the bolt length in the assembly to have a length of 1.25 inches by changing your Work Part to the component part. The Displayed Part can be the top-level assembly, a subassembly, or the bolt itself. Once the Work Part change is made, edit the length of the bolt. When the edit is complete, all 150 bolts update automatically.

Note: When you select **File** | **Save**, the current Work Part and ONLY the modified components underneath are saved. Therefore, if you save the top-level assembly, you automatically save ALL of the components.

Changing the Work Part:

Assemblies | Change Work

Use this to change the active part. The Work Part is the only part in the assembly that can have edits performed on it. The Work Part can be a subassembly part or an individual component.

Exercise: Changing the work part

1. Open **assy_1**.
2. Select **Assemblies | Change Work Part**.
3. Select a part from the **Choose loaded part** list or from the Graphics window. If a part is selected from the Graphics window, it highlights in the **Choose loaded part** list.
4. Click **OK** to change the work part.

Notes: The **Change Work Part** command gives you the ability to design in context and is available only in the Gateway and Modeling Applications.

If there are multiple occurrences of a component you are trying to make the Work Part, select the individual component to be highlighted in the Graphics window. This keeps Unigraphics from randomly selecting one to highlight.

Exercise: Changing the displayed part

1. Continue with **assy_1**.
2. Select the desired part to change the displayed part.
3. Pick the part from the list or in the Graphics window. If a part is selected in the Graphics window, it highlights in the list.
4. Click **OK** to change the displayed part.

Note: Displaying the assembly while working on an individual component can be extremely helpful.

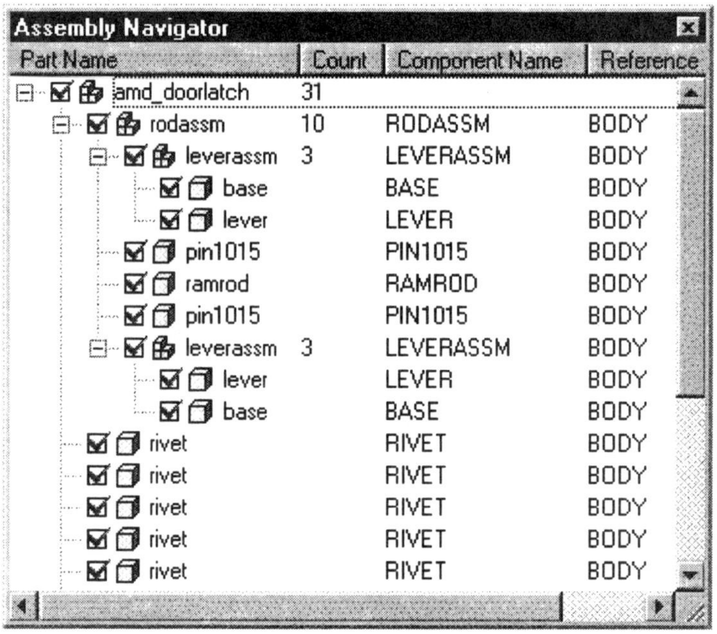

Assembly Navigator, (ANT):

The Assembly Navigator displays the assembly structure with graphical nodes allowing you to quickly determine the work part and invisible components.

Use **View | Assembly Navigator** to access the Assembly Navigator. The Assembly Navigator window contains a graphical representation of the assembly tree structure. If the structure is too large for this window, you can use the scroll bars to pan up, down, left and right to view the rest of the structure. You can also resize the window, or move it to a different location on the desktop.

Each component in the assembly tree displays as an entry, or *node*, in the Assembly Navigator. Use these nodes to select components as you would in the Graphics window, whenever you are prompted for a selection.

When accessing the Assembly Navigator for the first time, a message prompts you to configure the displayed columns, if desired. Select **Configure Now** to display the **ANT Column Configuration** dialog, allowing you to customize the Assembly Navigator. Refer to the Assembly Navigator Column Configuration section for more details on customization. **Accept Default** allows you to continue with the default configuration.

You can display a toolbar for the Assembly Navigator tools using **Tools | Customize | Toolbars**. However this selection is not available until you open the Assembly Navigator for the first time in each UG session.

The assembly structure displays as a tree. Each component, whether a piece part or sub-assembly, is indented. Each indentation represents a level in the structure.

Sort the structure by clicking the title of a column, click to alternate between an ascending and descending order on the column. The sort applies to each level independently. The top-level components are sorted first, and then the second level, continuing sequentially thereafter.

An icon in the **Part Name** column represents each of the components in the Assembly Navigator:

Represents apiece part.
Represents a sub-assembly.

When you place the cursor over a piece part node and click, the component highlights in both the Graphics window and the Assembly Navigator.

Either a positive or negative sign precedes the top assembly and subsequent sub-assemblies. Click "**+**" to expand the branch of the tree following the sign. Click "**-**" to collapse it. This does not remove the components from the assembly tree only from the tree display, reducing the amount of information displayed in the Assembly Navigator. Expanding and collapsing has no effect on the display of components in the Graphics window.

When the top assembly is the work part, all block icons are active, displaying in yellow. When a component is made the work part, only its respective icon and any sub-components are active. The rest of the block icons in the tree are inactive.

To make a component the work part, use the pop-up menu or double-click its icon.

A check box precedes each of the block icons. If the check box is empty, the component is unloaded. The block icons display in a gray dashed font:

☐ Represents an unloaded piece part.
☐ Represents an unloaded sub-assembly.

Click the check box to load the component. A check mark in the box means that the component is at least partially loaded. The block icons now display in gray solid font and are yellow in color:

☑ Represents apiece part at least partially loaded.
☑ Represents a sub-assembly at least partially loaded.

When a check mark displays, its color has meaning. A red check mark means the component's current reference set is visible, the component is unblanked and on a visible layer. Click a red check mark to make a component invisible by blanking it, represented by a gray check mark. This is the same operation as **Edit** | **Blank** | **Blank**, or selecting **Blank** from the Assembly Navigator pop-up menu.

☑ Represents a visible component.
☑ Represents an invisible component.
☐ Represents an unloaded component.

When a component is blanked, it is made invisible in the Graphics window, but is not removed from the assembly.

In the Assembly Navigator place the cursor over a blanked component, noting the red block representing the component with the part name displays in the Graphics window. This technique allows you to quickly locate blanked components. To use this technique select **Preferences | Assemblies** and activate, **Preselect Invisible Nodes**.

Notes: Press MB1 in the Assembly Navigator to check a component's location, which occurrence of a component is being chosen, and where thatoccurrence is located within the assembly.

The Assembly Navigator is probably the most effective tool for selecting components because it allows you to view the tree structure, and identify the current Work Part. Although you can select components in the Graphics window or the **Select Component** dialog, we recommend using the Assembly Navigator because it is the most informative.

Assembly Navigator Pop-up Menu:

The Assembly Navigator, along with its pop-up menu, is a powerful tool permitting easy navigation and control over an assembly structure. You can control the work and displayed parts, blank and unblank, change reference sets, gather component information, and locate the component in the assembly structure using this tool.

By placing the cursor over a component and right clicking, a pop-up menu displays, that allows you to perform operations on the assembly structure.

Some of these commands are shortcuts. For example, you can blank components by selecting **Blank** from the Assembly Navigator pop-up menu rather than using **Edit | Blank | Blank** from the main menu. Once a component is blanked, **Unblank** replaces **Blank** in the pop-up menu.

> **Make Work Part** is the same as selecting **Assemblies | Change Work Part** or double-clicking the node in the Assembly Navigator.
>
> **Make Displayed Part** is the same as selecting **Window | More Parts**. When you make a component the displayed part, only that part and its components display in the assembly tree and the Graphics window.
>
> **Display Parent** is only available when you pick a part that is a component of some other part (i.e., at least a first level sub-assembly). It displays a submenu listing the parents. The part you select becomes the new displayed part. All of the

components of the displayed part become visible in both the assembly tree and the Graphics window.

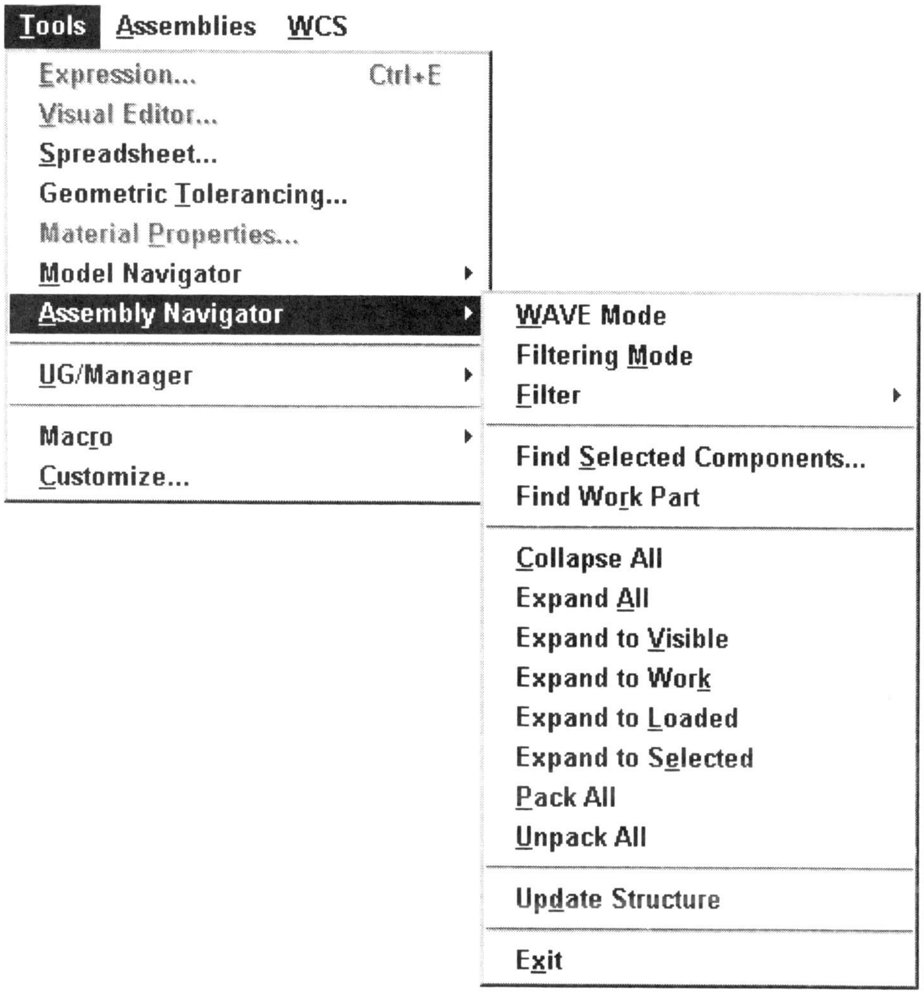

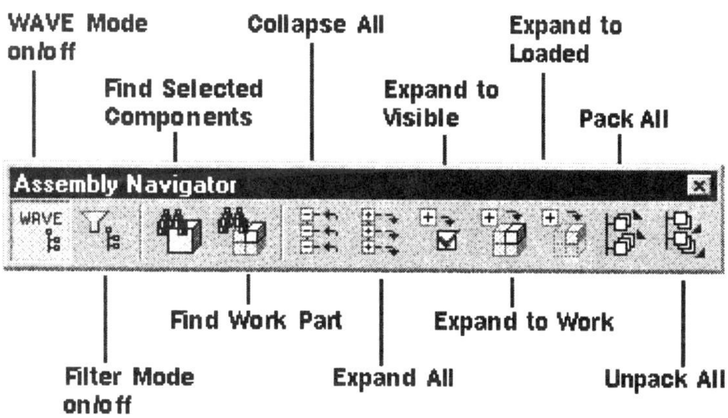

Bottom-Up Assembly:

Use this when the necessary components are already modeled- you just need to put them.

Assemblies | Edit Structure | Add

Multiple Add:

Use this list to efficiently add multiple components.

No creates a single component

Array allows you to create a linear or circular array of components.

Chain allows you to create components at any number of locations using the **Point Constructor** dialog

Positioning:

Use this list to specify the location of the component within the assembly.

Absolute will allow you to position a component using the **Point Constructor** dialog.

Mate allows you to assign an associative positioning relationship between the component you are adding and an existing component in the assembly.

Reposition allows you to move the component (translate, rotate, or reposition) after specifying an initial location.

Layer Options:

This list allows you to specify the assembly layer you want the component to display on.

Work places the component on the current work layer of the assembly.

Original places the component on the same layer as in the component part file.

As Specified allows you to specify the layer to place the component.

Exercise: Adding a component to an assembly with bottom-up techniques

1. Create an assembly file named **assembly_1**.
2. Select **Assemblies | Edit Structure**.
3. Click **Add** to display the **Select Part** dialog.
4. Select **base_part** to add to the assembly. The part can be selected by clicking **Choose Part File**.
5. Click **OK** to display the **Add Existing Part** dialog.
6. Select **Multiple add | No** from the list.

7. Select a **Positioning |Absolute** from the list.
8. Select **Original** as a **Layer options** setting.
9. Click **OK** to display the **Point Constructor** dialog. Select a point and click **OK** to add the component.

Notes: Each component inserted using **Add** is created as a component of the existing Work Part. If you have several subassembly levels, pay close attention to the Work Part.

You can specify different component names for the same part file. For example, if you are adding another occurrence of a part already in the assembly, you can specify a different component name to easily distinguish between the occurrences.

When using **Reposition** with multiple components, you position all occurrences simultaneously. To move an individual occurrence use **Assemblies | Edit Structure Reposition.**

Components can have a different color, font, width, and layer setting in the assembly file than in the components part file.

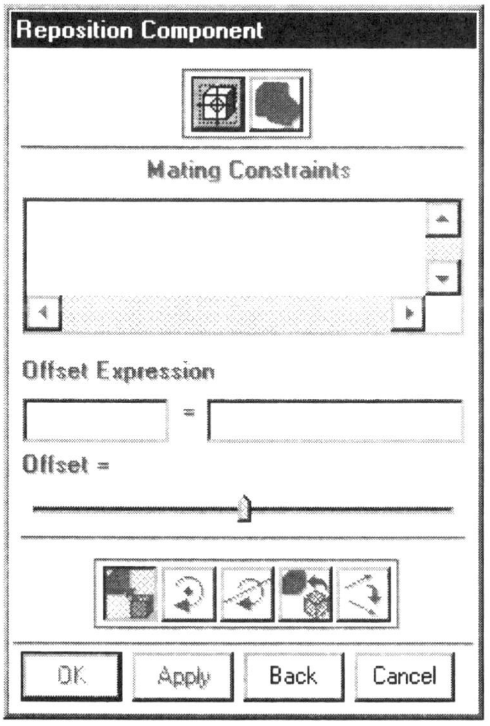

Repositioning Components:

Assemblies | Edit Structure | Reposition

Exercise: Dragging a component

1. Open **assy_2.prt**.
2. Select **Assemblies | Edit Structure**.
3. Click **Reposition** to display the **Reposition** dialog.
4. Pick a component in the Graphics window or Assembly Navigator. Only one component can be repositioned with this method at a time.
5.  Click **Finished Selection**.
6. The standard crosshairs change to crosshairs and a dot.
7. Click in the Graphics window and drag the component to a new location.
8. Click **OK** to accept the new location.

Exercise: Translating components

1. Continue with **assy_2**.
2. Select **Assemblies | Edit Structure**.
3. Click **Reposition**.
4. Pick a component in the Graphics window or Assembly Navigator. Multiple components can be selected.
5.  Click **Finished Selection**.
6.  Click **Translate**.
7. Click the desired transform method and define the parameters.
8. Click **OK** to return to the **Reposition Component** dialog and translate the component.
9. Click **Cancel**.

Exercise: Rotating components about a point

1. Continue with **assy_2**.
2. Select **Assemblies | Edit Structure**.
3. Click **Reposition** to display the **Reposition** dialog.
4. Pick a component in the Graphics window or Assembly Navigator. Multiple components may be selected.
5.  Click **Finished Selection**.
6.  Click **Rotate About a Point**.
7. Pick a rotation center point.
8. Enter the **angle** in the **Angle** text box. The angle is measured about ZC-axis.
9. Click **OK** to return to the **Reposition Component** dialog and rotate the component in the XC-YC plane.
10. Click **Cancel**.

Exercise: Rotating components about a line

1. Continue with **assy_2**.
2. Select **Assemblies | Edit Structure**.
3. Click **Reposition** to display the **Reposition** dialog.
4. Pick a component from the Graphics window or Assembly Navigator. Multiple components may be selected.
5. Click **Finished Selection**.
6. Click **Rotate About a Line**.
7. Click the desired transform method and define the parameters.
8. Click **OK** to return to the **Reposition Component** dialog and rotate the component.
9. Click **Cancel**.

Exercise: Rotating components between axes

1. Continue with **assy_2**.
2. Select **Assemblies | Edit Structure**.
3. Click **Reposition** to display the **Reposition** dialog.
4. Pick a component from the Graphics window or Assembly Navigator. Multiple components may be selected.
5. Click **Finished Selection**.
6. Click **Rotate Between Axes** to display the **Point Constructor** dialog.
7. Pick a reference point. The Vector Constructor dialog displays.
8. Define a vector and click **OK**.
9. Define a vector and click **OK**.
10. Enter an angle in the **Minor Angle** text box. By default, the angle between the two defined vectors is entered. Click **Major Angle** to display the complement angle in this box.
11. Click **OK** to return to the **Reposition Component** dialog and rotate the component in the plane defined by the two vectors.
12. Click **Cancel**.

Note: When using **Reposition** on partially constrained components, first activate **Show Degrees of Freedom** from the mating condition pop-up menu in order to understand the allowed movement.

Mating Conditions:

There are two methods for positioning components within the assembly structure: absolute and mating conditions. The absolute method is not associative. For example, if a bolt, nut, and washer are aligned using this method and the bolt is then moved, the alignment relationship is not maintained and the assembly structure may nolonger be valid. The Mating Conditions method allows you to establish positioning relationships between components in the assembly structure. These relationships enable you to link components so that modifying or repositioning one component in the assemblystructure also updates any mated components, maintaining the established relationships.

Mating conditions allow you to define an associative positioning relationship between components. These relationships are established by creating constraints between two components in the assembly: the base component and the mated component. Once you establish these relationships, an associative positioning relationship between the base and mated components is maintained. Using the Mating Conditions method of positioning, you could set up a relationship between the bolt, nut, and washer that keeps them coaxial. With this relationship established, the nut and washer move when the bolt is moved.

Mating conditions can be created when components are added to the assembly or between components previously added using the absolute positioning method.

A Mating Constraint is a positioning relationship between two components.

A Mating Condition is a set of mating constraints for a particular component.

A Mated Component is a component to be positioned using a mating condition.

A Base Component is a component with positioning relationships assigned between it and the mated component. The base component is considered the**To** component.

A Surface Normal is a vector that is perpendicular to a face or plane at any given location on the face or plane, pointing away from the part.

An Axial Normal is a vector that is collinear to the center axis of a cylindrical face.

The **Mating Conditions** dialog presents the mating conditions and constraints in a tree structure format. Conditions and constraints are created and edited from this structure using a set of pop-up menus.

The list at the top of the **Mating Conditions** dialog displays a tree structure similar to that of the Assembly Navigator. The first entry in the structure, known as the root node, represents the assembly in which you are creating the mating conditions. The level under it represents the mating conditions. Preceding each of the mating condition nodes is the plus or minus signs to collapse ("**-**") and expand ("**+**") the branch of the tree following the sign.

The check box preceding the mating condition node controls the suppression of the mating condition:

☐ Represents a suppressed mating condition.

☑ Represents an unsuppressed mating condition.

When the structure is expanded, the next level represents the mating constraints.

Creating Mating Conditions:

Create mating conditions by using **Assemblies | Edit Structure | Mating Conditions**. Click this button to edit existing mating conditions or create new ones. You can still access the **Mating Conditions** dialog when adding new components using **Add**.

Once the **Mating Conditions** dialog displays, create a mating condition by either selecting a mating type and component to mate.

Mating Constraints:

The three-dimensional constraint solver supports eight constraint types, offering great flexibility.

There are eight constraint types. After constraints are created, the mated component repositions based on its constraints. Partially constrained mated components do not necessarily reposition next to the base component. Likewise, after constraints are deleted, the mated component does not reposition.

The constraint types are:

Mate

This constraint type positions two objects of the same type so that they are coincident. When you mate two faces, the surface normals of the two faces point toward one another and place one face on top of the other.

Align

This constraint type positions planar objects so they are coplanar. It positions cylindrical, toroidal, and conical faces so they are aligned along their axes and it positions edges to be collinear.

Angle

This constraint type allows you to define an angle between two objects. In addition to the direction vectors, an angle axis displays when **Highlight | With Direction** is selected from the mating condition pop-up menu. The angle axis is perpendicular to both direction vectors and defines the direction of rotation. If the angle axis is pointing toward you, the angle is measured in a counterclockwise direction between the direction vectors.

There are two types of **Angle** constraints, **Planar** and **3D**. The **Planar** method requires you to define the angle axis, which must be perpendicular to both direction vectors. With **3D** the angle axis is defined for you.

Parallel

Select this icon to establish parallelism. The direction vectors of the mated and base components are aligned, but the distance between the objects is not considered.

Perpendicular

Use this constraint to establish the direction vectors of the mated and base components positioned at a 90-degree angle to one another, but the distance between the objects is not considered.

Center

This type of constraint positions the mated component based on the selected objects to center. One or two objects may be used on the mated component, base component, or both. Use the **Center Objects** list to specify this. The first number in the name of each option refers to the mated component while the second refers to the base component:

1 to 1 means 1 mated object centers to 1 base object.
1 to 2 means 1 mated object centers to 2 base objects.
2 to 1 means 2 mated objects center to 1 base object.
2 to 2 means 2 mated objects center to 2 base objects.

Distance

This constraint function is essentially the same as **Mate** or **Align**, except an offset value can be specified. In this example, the direction vectors are pointing toward each other, the same as a **Mate** constraint.

Click **Alternate Solution** to reverse the direction vectors. In this example, the direction vectors are pointing away from each other, as with an **Align** constraint.

Tangent

Use this constraint to define a tangency between two objects. Here are some examples:

- A point, line, or planar face tangent to a cylindrical face or a spherical face.
- A cylindrical or spherical face tangent to another cylindrical or spherical face.

Selection Steps:

These icons help you navigate through the mating condition. Use two, three or all four steps, depending on the type of constraint you are creating or edting. The steps alternate between the mated and base components. As you define your constraint, you move to the next step automatically when making a selection. To view previous selections click the desired icon. This eliminates the need for a **Back** button, which is no longer available in the **Mating Conditions** dialog. In each of the steps, **From** refers to an object on the mated component. **To** refers to an object on the base component.

 **From**

 **To**

 **Second From**

 **Second To**

Notes: The **Highlight | With Direction** command in the mating constraint pop-up menu is an excellent tool to help you visualize possible mating constraint solutions.

The **Angle** constraint works best if it is used last.

When viewing highlighted faces and direction vectors, select a specifc constraint from the tree display to view the faces and vectors for that constraint only.

You can add only one component at a time when **Positioning** is set to **Mate** in the **Add Existing Part** dialog.

Assemblies | Edit Structure | Mating Conditions

Exercise: Creating mating conditions

1. Open **assy_3**.
2. Create the constraint (use the desired process).
3. Review the degrees of freedom remaining.
4. Create additional constraints until you have the desired mating condition.

Exercise: Creating a mate, align and tangent constraint

1. Continue with **assy_3**.

2. Select the desired icon in the **Mating Type** group box of the **Mating Conditions** dialog.
3. Pick the object; such as a face or edge, on the component you are mating from (the mated component).
4. Pick the object on the component you are mating to (the base component).
5. Click **Apply** or **OK** to accept the mating condition.

Exercise: Creating an angle constraint

1.  Click **Angle**.
2. Select the desired **Angle** from the list, **Planar** or **3D**.
3. Pick the object on the mated component.
4. Pick the object on the base component.
5. Pick the object to define the axis, only applicable for **Planar Angle**.
6. Enter the expression name and angle value in the **Angle Expression** text boxes.
7. Click **Apply** or **OK** to accept the mating condition.

Exercise: Creating a parallel and perpendicular constraint

1. Click the desired **Mating Type**.
2.  **Parallel**
3.  **Perpendicular**
4. Pick the object, such as a planar face or linear edge, on the component to be mated from (the mated component).
5. Pick the object on the component to be mated to (the base component).
6. Click **Apply** or **OK** to accept the mating condition.

Exercise: Creating a center constraint

1.  Click **Center**.
2. Select the desired **Center Objects** setting, such as **1 to 1**.
3. Pick the object on the component to be mated from (the mated component).
4. Pick the object on the component to be mated to (the base component).
5. If **Center Objects** is set to **2 to 1** or **2 to 2**, pick the second object on the component to be mated from (the mated component).
6. If **Center Objects** is set to **1 to 2** or **2 to 2**, pick the second object on the component to be mated to (the base component).
7. Click **Apply** or **OK** to accept the mating condition.

Exercise: Creating a distance constraint

1.  Click **Distance**.

2. Pick the object on the component to be mated from (the mated component).
3. Pick the object on the component to be mated to (the base component).
4. Enter the expression name and offset value, if desired, in the **Offset Expression** text boxes.
5. Click **Alternate Solution** to switch between a mated offset and an aligned offset constraint.
6. Click **Apply** or **OK** to accept the mating condition.

Exercise: Editing a mating constraint

1. Click the mating constraint to edit from the tree structure at the top of the **Mating Condition** dialog.
2. Click the **Selection Step** to edit, if desired, and pick the new object.
3. Click **Apply** or **OK** to accept the mating condition.

Notes: When you want to mate a component while adding it to an assembly, set **Positioning** to **Mate** in the **Add Existing Part** dialog.

When creating a **Distance** constraint, you can use negative values for the **Offset Expression** to indicate an offset in the opposite direction.

Apply **Alternate Solution** to a **Distance** constraint to flip the direction the component is facing.

Use **Preview** to visualize the way mating constraints will be solved. **Unpreview** allows you to undo without repositioning, making this a great tool for learning the new constraint types.

Possible Mating Condition Problems:

When you establish the first mating constraint, Unigraphics checks to confirm the constraint is valid. If it is a valid constraint, the status line tells you how many degrees of freedom remain for the mated component. The degrees of freedom canbe displayed in the Graphics window and change depending on how many remain. There are four degrees of freedom that you can constrain: three linear and one rotational.

To completely constrain a mating condition, create mating constraints to eliminate all degrees of freedom. You do not have to fully constrain a mating condition to establish an associative positioning relationship.

However, if you do not fully constrain a mating condition, it is possible that there is more than one way to solve the constraints you have applied.

If the solver cannot solve the mating constraint, the Status line displays the message "Conflicting constraints found".

Notes: To remove linear degrees of freedom, use the **Mate**, **Align, Distance, Center,** or **Tangent** options. To remove a rotational degree of freedom, use the **Angle**, **Parallel**, or **Perpendicular** options.

It is almost always easier to create the mating condition when you use the **Mate**, **Align**, or **Distance** constraints first.

Top-Down Assemblies:

Assemblies | Edit Structure | Create

There are many advantages to Top-Down modeling, the greatest being the ability to design or edit in the context of the assembly structure. It is important to understand the following concepts:

- Define the new component before creating any parametric geometry when possible.
- Some applications and operations are independent of the Work Part.
- Sketching in the context of an assembly.
- The role of the Work Part in defining the assembly structure.
- Switching back and forth between Top-Down and Bottom-Up methods.
- Creating subassembly files.

Before creating any parametric geometry for a new component, you should define the component using **Assemblies | Edit Structure | Create** and then set the Work Part to be the new component. If you fail to complete these steps prior to creating new geometry, you are creating the new geometry in the current Work Part. Each unique component has its own unique part file, which contains its own geometry. To maintain the organization of the assembly you must move the new geometry into its own part file.

When using the design-in-context approach, be aware of the work part. All new components are created in the current work part. All component manipulation such as repositioning and removing can only be performed on components that are one level below the current work part. All geometry creation and editing is done in the work part.

Exercise: Creating an assembly using top-down methods

1. Select **File | New**.
2. Enter a name in the **File name** text box.
3. Click **OK** to create the assembly part.

Exercise: Creating a new component using top-down methods

1. Use the above file.
2. Select **Assemblies | Edit Structure**.
3. Click **Create**.
4. If desired, select any objects to include in the component. The following objects can be selected: curves, sketches, solid bodies, sheet bodies, coordinate systems, datum planes, datum axes, groups, components. A new component can also be created with no initial geometry.
5. Click **OK**.
6. Enter a name for the new component in the **File name** text box.
7. Click **OK** to create the component part and display the **Create New Component** dialog.
8. If desired, enter a component name in the **Component name** text box. By default, the part name is already entered.
9. Select a **Component Origin** setting.
10. Click **OK** to create the component in the assembly. The component is created at the origin of the coordinate system chosen in the **Component Origin** option. Watch the Status line for an indication that the component is created.
11. Click **Cancel**.

Notice the assembly you created is the work part. Select **View | Assembly Navigator** to view the structure. Change the work part to the component you created to model the component.

Combining Top-Down and Bottom-Up Methods:

You can use a combination of Top-Down and Bottom-Up methods in defining the assembly structure. You can also use a combination of these methods in designing and editing individual component geometry. For example, in an existing assembly, you create a new component and a "black box" shape to define its location and overall size within the assembly structure. You can save this part, close the assembly, and design the detail of the component part using Bottom-Up methods. The combination of methods provides the best of both worlds: Top-Down for positioning and evaluating size restrictions and Bottom-Up for detailing the component without having to work with the entire assembly.

Creating Sub-Assemblies:

Remember, components are the piece parts making-up an assembly. These piece parts can be used in multiple assemblies and multiple sub-assemblies. Therefore, it is always a good idea for each unique piece part to have its own part file. Even these piece parts can have their own component piece parts. When you add a sub-assembly to an existing assembly, all of the components are added as well. In terms of design-in-context concepts, use the **Assemblies | Edit Structure | Create** to create sub-assembly files before adding or designing new components. This way you create a sub-assembly structure that is much easier to manage and manipulate than if you simply add the components to the top-level assembly.

Component Loading:

Components by default are only "partially loaded" when an assembly is loaded, decreasing the amount of time required needed to load the assembly and all of the components. Only the amount of data required to display the component in the assembly structure is loaded. Select **List Components** (**Assemblies | Reports**) to check which components are partially loaded.

Note each component is labeled **Partially loaded** in the Comment column. When a component is the work part, Unigraphics loads all of the information required and removes the column comment.

Load Options:

When you open an assembly file, the system needs to find the components that make up the assembly. The component part files making up an assembly can exist in many different locations within your system. Unigraphics needs to know where to search for these components. This option is controlled by the **Load Options** dialog which displays when you select **File | Options | Load Options**.

Load Components:

The components to load when an assembly file is loaded are defined by the **Load Components** list.

All Components allows you to load all of the components in any assembly.

No Components allows you to load the assembly file but not the components. The components are all referred to as **Not loaded**. The advantage of this approach is that you can load large assemblies quickly and then determine the components to open using the Assembly Navigator, the system-defined Component Filters, or the user-defined Component Filters. Component Filters open all the components specified in the filter.

Use Last Component Set allows you to load the same set of components that were loaded when the assembly was last saved.

Use Last Filter allows you to load the set of components that match the last filter used to load the assembly.

Specify Filter the set of components that match a filter selected from a listing when the assembly was loaded.

This allows you to partially load the components of an assembly. This can save virtual memory (20% to 30% and higher according to tests) resulting in significant performance improvements. You should take advantage of partial loading in most situations.

Deactivating **Use Partial Loading** allows you to fully load the components and can be useful in the following situations:

- Saving an entire assembly in a new version of Unigraphics. Only fully loaded component parts are saved when you save the parent assembly.
- Making changes to assemblies in which you desire to see instant updating of interpart expressions.

Abort Load on failure:

If Unigraphics cannot load the components in an assembly, either because they could not be found based on the current Load options or because they are no longer on the system (i.e., deleted or archived), the loading of the assembly must either be aborted or modified. The **Abort Load on Failure** option controls this. When components cannot be located while the assembly is loaded, the **Open Warning** message appears. When this option is active, the load aborts.

When this inactive, Unigraphics loads the assembly and any components that can be located based on the current Load options. The remaining components that could not be located are not loaded and are referred to as "Not loaded." If the components are subsequently located, they can be opened using **Open | Component** on the Assembly Navigator pop-up menu. The assembly is then updated with no loss of associativity.

Allow Substitution:

This provides the capability to load components having the same name as the original component but not the same internal identifier. These parts are not related to one another in any fashion except for the part file name. One of these parts is added to an assembly as a component. Later that cover is deleted from the operating system and it is desired to use the second cover in the assembly. Even though the second **cover.prt** part file could be located, it would not load it into the assembly because the two parts are not the same internally.

If **Allow Substitution** is active, Unigraphics exchanges the second **cover.prt** part file for the first. A report is issued indicating that a substitution has taken place. You need to understand that because internally these two covers are different components, all associativity to the original plate is lost when the second plate is substituted. This option is very similar to using **Assemblies | Edit Structure | Substitute**.

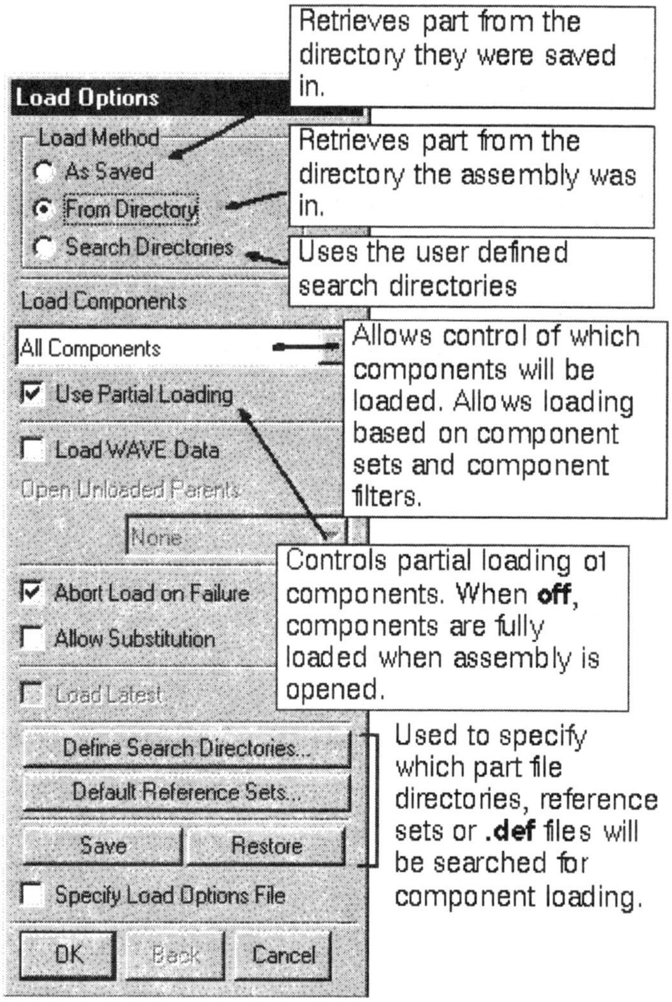

Load Method:

The load method determines where the system will search for the components. There are three possible settings:

As Saved looks for each component in the same directory that the component part was saved in.

From Directory looks for each component in the directory that you are currently using.

Search Directories looks for each component in the directory(s) that is specified in the **Define Search Directories** dialog area.

Default Reference Sets:

This allows you to define the default reference sets to use when components are loaded. When a component loads, it loads the first reference set matching this list. The default reference sets **As Saved**, **Entire Part** and **Empty** already display in the list.

Apply to All Levels:

This changes the way an assembly is loaded. With the **Apply to All Levels** option active, all occurrences of the first default reference set are loaded. With this option inactive, after a subassembly is found that does not have the first default reference set in it, none of the components that make up that subassembly loads including that first default reference set. This is true even if the reference set is first on the **Preferred Reference Sets** list.

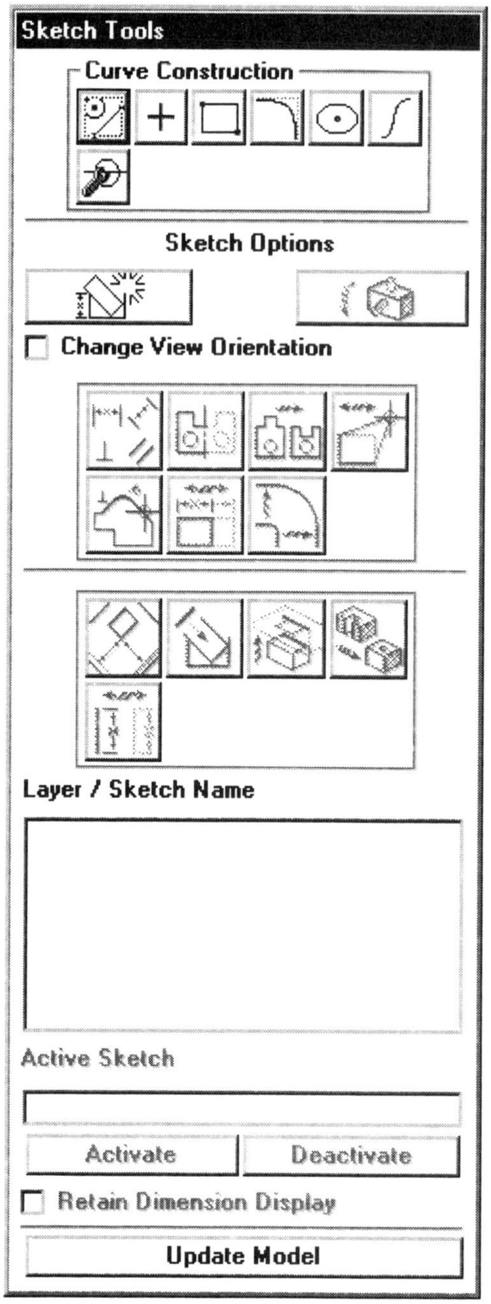

Sketching in an Assembly:

 **Insert | Sketch**

Create sketches in a work part while having the assembly as the displayed part. As you create sketches, you can reference geometry from the displayed part for measuring and extracting curves. This technique speeds up the design process.

Exercise: Creating a sketch in an assembly

1. Open **assy_2**.
2. Select **Assemblies | Change Work Part**.
3. Select a component.
4. Click **OK**.
5. Change the work layer to the desired layer for the new sketch.
6. Select **Insert | Sketch** to display the **Sketch Tools** dialog.
7. 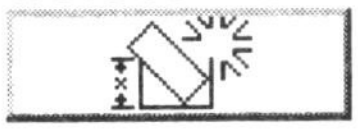 Click **Create** to display the **Create Sketch** dialog.
8. Enter a name in the **Sketch Name** text box.
9. Pick a planar face or datum plane as the attachment face/plane. You can choose only objects from the work part. Use a plane of the WCS by selecting **XC-YC**, **XC-ZC**, or **YC-ZC** from the **Sketch Plane** list.
10.  Click **Reference Direction** or MB2-click.
11. Pick an object to define the horizontal reference. If none is available, the **Reference Direction** can be set to **Vertical** and an object picked to define the vertical reference. A direction vector displays.
12. If the vector is not pointing in the desired direction, click **Reverse Direction**.
13. Click **OK** to return to the **Sketch Tools** dialog.
14. Create the sketch curves and constraints.
15.  If desired, click **Positioning Dimension** and position the sketch.
16. Click **Deactivate**.

Notes: When the assembly is the work part, you can only position a sketch using the WCS of the displayed part as a reference.

Removing Components from an Assembly:

Assemblies | Edit Structure | Remove

Exercise: Removing a component from an assembly

1. Change your Work Part to the immediate parent of the component that is to be removed.
2. Select **Assemblies | Edit Structure** to display the **Edit Assembly Structure** dialog.
3. Click **Remove** to display the **Choose Component** dialog.
4. Select the component you want to remove from the list or enter the name in the **Component name** text box to remove all occurrences of the component. Select a component from the Graphics window or Assembly Navigator to remove a particular occurrence of the component.
5. Click **OK**.

Notes: When I select **Assemblies | Edit Structure | Remove,** the **No Valid Components** dialog appears. Why can't I select components to remove? Your Work Part is not the assembly. Check your current work part labeled in the title bar of the Graphics window and use **Assemblies | Change Work Part** to make the assembly the current work part.

When choosing multiple components from the component list, you can select them at the same time, just click and drag the cursor down the list. If the components to be selected are not in order in the list, pick the first component and then select the remaining components while pressing the CTRL key. To select an inclusive list of components, select the first component and select the last component while pressing the SHIFT key.

There are several methods for selecting and deselecting components in the Graphics window. You can drag a rectangle to select components. You can deselect components by pressing SHIFT while selecting the highlighted components. You can deselect the last component while selecting a new component by pressing the CTRL and SHIFT keys.

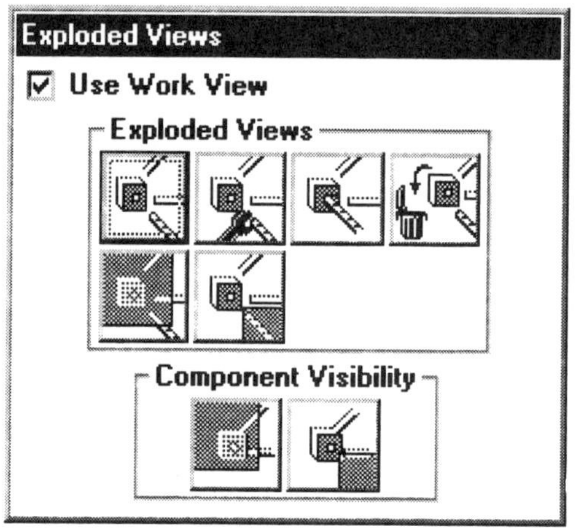

Exploded Views:

There are three different methods of creating an exploded view.

This will explode the components for you. This explosion is intended for use as a starting point for mated components. This means that once the automatic explosion is complete, you probably need to edit the explosion in order to move the components to their final exploded location.

This method allows you to explode all of the selected componentsas a group. For example, if a sub-assembly is picked, the components cannot explode independently of the sub-assembly. This method is useful when you wish to move the selected components or entire sub-assemblies as a single object.

This method applies different explosion parameters to each of the selected components. For example, if a sub-assembly is picked, you can explode the sub-assembly as well as each of the components within the sub-assembly. This method is useful when you wish to apply different explosion parameters to the selected components.

When an exploded view is created, a new view name is created by default and the explosion is view dependent in that view unless you specifically request it in another view. If **Use Work View** is active in the **Exploded Views** dialog, the explosion is created in the work view. If you are working in a multi-view layout, you can deactivate this and Unigraphics then allows you to pick a view in which to create the explosion. The selected view has the word EXPLODED appended to its name (i.e., TOP becomes TOP (EXPLODED)).

Notes: You may want to create a unique view using **View | Operation | Save As** before creating any explosions. This results in better organization of your assembly files and easier view management in the Drafting application.

Whenever you see EXPLODED next to a view name, you have to understand the components may or may not have been exploded. Select **Assemblies | Exploded Views | Hide Explosion From View** to ensure the components are in their true positions.

Use **Individually** as the **Explode** method if you want to explode an entire sub-assembly and then explode the components subsequently. Unigraphics prompts you to explode the sub-assembly first, then each of the components in the sub-assembly.

Creating an Explosion:

Assemblies | Exploded Views | Create Explosion

Revert to Real Position allows you to move an exploded component back to its true model position within the assembly. In addition, while this is active, the component cannot have any other explosions applied.

Apply General Transform allows you to explode the component using the same set of general transform methods as using **Assemblies | Edit Structure | Reposition**.

Explosion Vector will determine the direction and distance in which to explode the selected component.

> **From Mating Condition** is only available for mated components and inherits the direction vector from the mating condition.
>
> **I / J / K Value** defines the direction vector relative to the current WCS.
>
> **Distance** allows you to enter a magnitude for the explosion. The distance you enter applies along the explosion vector direction.

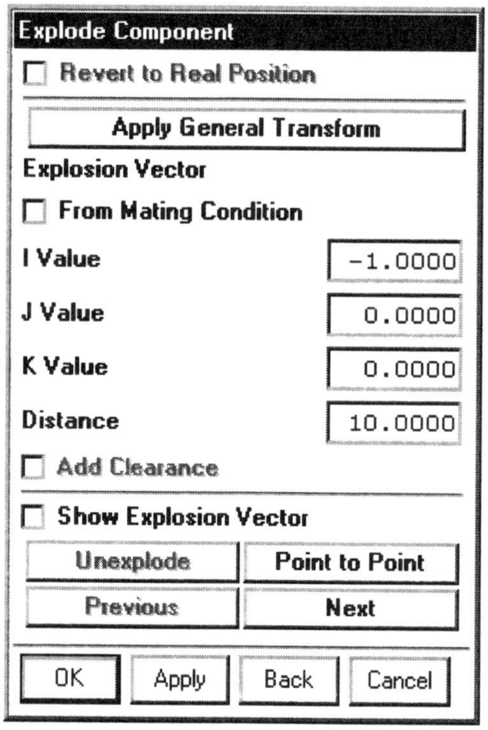

Exercise: Creating an exploded view automatically

1. Open **assy_5**.
2. Select **Assemblies | Exploded Views**.
3.  Click **Create Explosion**.
4. Enter the name of the explosion in the **Name** text box. By default, the explosion name is numbered sequentially starting with Explosion 1.
5. Click **OK**.
6. Pick the components to explode from the list, the Graphics window or the Assembly Navigator. Only components one level below the work part can be picked.
7. Set **Explode** to **Automatic**.
8. Click **OK** to display the **Explosion Distance**. All components with mating constraints are exploded. Since it is difficult to predict the desired offset, it is probable that you will need to edit the explosion upon completon. Any components without mating constraints are ignored. If none of the components are mated, the **Information** message appears.
9. Click **Cancel**.

Exercise: Creating an exploded view together or individually

1. Select **Assemblies | Exploded Views** to display the **Exploded Views** dialog.

2. 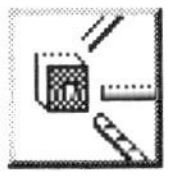 Click **Create Explosion**.
3. Enter the name of the explosion in the **Name** text box.
4. Click **OK** to display the **Select Components** dialog.
5. Pick the components to explode from the list, the Graphics window or the Assembly Navigator. Only components one level below the work part can be picked.
6. Set **Explode** to either **Together** or **Individually**.
7. Click **OK**.
8. Indicate the direction and distance by entering values in the **I Value**, **J Value**, **K Value**, and **Distance** text boxes.
   Or
   Click **Apply General Transform** to use a general transform method.
9. If desired, activate **Show Explosion Vector** to verify the explosion direction of the component.
10. Click **Apply** to the explosion vector and distance to the highlighted component or group of components.
11. Click **Next** to highlight the next component or group of components.
12. Repeat Steps 8, 9, and 10 for each component or group of components.
13. When all of the components are exploded, the **Next** button is no longer available. Click **OK** to return to the **Selected Components** dialog.
14. Click **Cancel**.

Notes: You can use **Apply General Transform** regardless of whether or not the component has a mating condition associated with it.

When **From Mating Condition** is used, **Point-to-Point** only determines the magnitude of the explosion along the explosion vector.

Reference Sets:

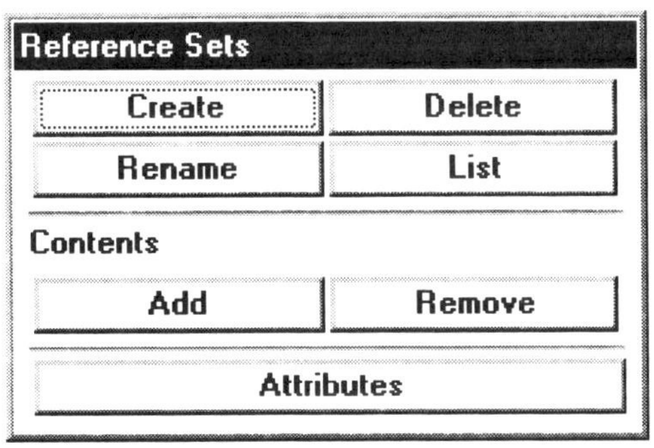

Assemblies | Reference Sets

Create creates a reference set. The name you specify should be descriptive of what the reference set contains.

Create Ref Set CSYS - No
Use to specify an orientation for the reference set. Otherwise, the Absolute Coordinate System is used for orientation.

Delete allows you to delete a reference set by choosing the reference set name from a list and confirming your wish to delete the selected reference set. The geometry contained in the reference set is not deleted, only the reference set.

Rename will rename an existing reference set by selecting the desired reference set to rename from the list and entering a new name.

Add allows you to add objects to an existing reference set.

Remove allows you to remove objects from an existing reference set. You select a reference set and select objects to remove in the same manner used for the**Add** button.

Exercise: Creating a reference set

1. Open **assy_1**.
2. Select **Assemblies | Reference Sets**.
3. Click **Create**.
4. Enter a name for the reference set in the **Name** text box.
5. Click **OK**.
6. Pick the objects in the Graphics window to include in the reference set. You can add geometry, coordinate systems, datum planes, or any other entity type from the Work Part and its immediate components.
7. Click **OK**.
8. Click **Cancel**.

Exercise: Adding objects to a reference set

1. Select **Assemblies | Reference Sets** to display the **Reference Sets** dialog.
2. Click **Add** to display the **Choose Name** dialog.
3. Pick the reference set from the list, or enter the name in the **Ref. Set name** text box.
4. Click **OK** to display the **Class Selection** dialog.
5. Pick the objects in the Graphics window to include in the reference set. You can add geometry, coordinate systems, and datum planes from the Work Part and its immediate components.
6. Click **OK** to return to the **Create Reference Set** dialog.
7. Select **Cancel**.

Exercise: Replacing a reference set

1. Select **Assemblies | Edit Structure** to display the **Edit Assembly Structure** dialog.

2. Click **Replace Reference Set** to display the **Choose Component** dialog.
3. Pick one or more components from the list, the Graphics window, or the Assembly Navigator.
4. Click **OK** to display the **Choose Name** dialog.
5. Pick a reference set from the list.
6. Click **OK** to return to the **Choose Component** dialog.
7. Click **Cancel**.

The Master Model Concept:

The Master Model concept involves having a single definition of each piece part in an assembly. The single product definition allows each group within your organization to access the latest version of each piece part because it only exists in one location and in one part file.

The Assemblies application is the cornerstone of the Master Model concept. Each group within your organization works from a single part file for any particular component. However, concurrent engineering practices demand that groups have access to models during the design process. Even vendors require access to the model before the design is complete. The figure illustrates how assemblies enable the Master Model concept to work.

Each group creates an assembly file and adds the Master Model to the assembly file as a component. Each of the groups' work is performed in the assembly file without needing write access to the master model. For example, the NC data resides in the Manufacturing assembly file and the drawings reside in the drafting assembly file.

When the master model changes, the changes reflect in each group's assembly file automatically. Completely new versions of the master model can be created and reflected in each of the different assemblies.

Implementation of the Master Model concept requires a great deal of process planning as well as data management. It is good for you to be exposed to this concept and to research the feasibility of the Master Model concept in your own company.

Assembly Model Information:

The **Information** and **Analysis** menus allow you to gather information about the assembly structure. Information on objects, parts, distances and analysis functions are available. The commands work on any objects in the displayed part.An example of using this option is if a single component in an assembly is the work part and the entire assembly is the displayed part. If you perform a mass analysis on all of the objects, you calculate mass properties for the objects in the displayed part not just the work part.

Object Information:

Use **Information | Object**, to display information about an object and its owning part in an assembly.

Distance Information:

Use **Analysis | Distance** to measure the distances between object and components.

Displayed Part Information:

Use **Information | Part | Loaded Parts** to display information about all the loaded parts, including the work and displayed parts.

Reference Set Information:

Use **Information | Assemblies | Reference Set** to display information about reference sets including the contents of each reference set.

Mating Conditions Information:

Use **Information | Assemblies | Mating Conditions** to display a condensed version of the **Mating Conditions** dialog, listing the current mating conditions in the work part. The mating condition dialog is available to highlight the mating objects and show the degrees of freedom.

Advanced Assembly techniques:

Interpart Expressions:

It is possible to directly relate expressions between component parts. For example,a bolt exists in a unique component part file and a hole through which the bolt passes exists in another unique component part file. If the size of the hole changes, it is possible to automatically change the pitch diameter of the bolt shaft. This is accomplished by linking the parameter of the hole diameter to the parameter for the pitch diameter of the bolt shaft. This concept of linking expressions between part files is known as Interpart Expressions.

Expression Format:

The standard format for an interpart expression is:

[expression] = [linked part]::[expression from linked part]

Creating Interpart Expressions:

Tools | Expression | Create Link

Exercise: Creating a link

1. Open **assy_6**.
2. Select **Toolbox | Expression**.
3. Select an expression from the list to edit or create a new expression by entering it in the text box.
4. When an interpart expression is desired, select **Create Link** to display the **Select Part** dialog.
5. Select the component to link by selecting from the **Choose loaded part** list. Use **Choose Part File** to select unloaded components.
6. Click **OK** to display the **Expression List** dialog, listing the expressions of the selected part.
7. Select the expression from the list.
8. If the selected part is in different units, **Convert units** dialog box becomes available. Check the box to convert the units.
9. Click **OK** to return to the **Edit Expressions** dialog.
10. Finish editing or creating the expression and press ENTER.
11. Click **Apply** to apply the expression. Click **OK** to apply the expression and close the **Edit Expressions** dialog.

WAVE Geometry Linker:

The process of linking geometry from one part file to another is straightforward. You can perform many different types of design cases using WAVE techniques including concept design, assembly design, in-process models, and tooling design. When working with linked geometry and the features associated with it, you must become familiar with the way these functions work, modification methods, and manage techniques.

There are several methods you can use to link geometry:

Down the assembly structure, a sketch located in the assembly file is linked downward into the component parts to facilitate its design.

Up the assembly structure, curves are linked from the component part upward into the assembly file to form the beginning of a tooling fixture used in machining.

Across an assembly structure, a point is linked from a component in one subassembly to a component in another subassembly to locate a fastening hole. The hole is used to connect the two sub-assemblies together.

It is also important to understand that after you link the geometry from one part to another, you can reference it by subsequent modeling operations. For example, a solid body can be linked from one part file to a new component and then additional operations can be performed on that newly linked geometry.

Linked geometry can be used regardless of whether the original geometry is present at the time of use.

Links:

When you copy geometry from one part to another using the WAVE Geometry Linker, it is considered linked. As long as you maintain the link, the geometry updates. Broken links occur when:

- The parent part containing the original geometry is deleted.
- The path to the parent is removed through deletion or non-associative substitution.
- The parent geometry no longer exists.
- The link is deliberately broken.

Unless the link is deliberately broken, it is referred to as newly broken until you accept it. When you accept the newly broken link, it is considered broken until a new parent is defined. It is important to note that once broken, you can recreate links with new parents.

 When you select **Point**, the **WAVE Geometry Linker** dialog displays two **Point/Line** options to create linked geometry

 Select this to create points in the work part by extracting from existing objects. You can use the **Point type** list to select specific point types. Types include **Existing Point**, **End Point**, **Control Point**, **Arc/Ellipse/Sphere Center**, and **Quadrant Point**.

 Select this to create a series of lines between selected points. When points in the original part change, the line endpoints in the work part update accordingly (when associative links are created).

 Use **Curve** to extract an associative copy of a solid edge or two-dimensional curves.

 Use **Sketch/String** to extract associative copies of sketches or strings of curves from the component part. You can then use these sketches or strings to sweep new geometry or to position features in the work part.

In addition to selecting curves, you can select faces. When you select a face, Unigraphics creates a set of connected curves along each edge of the face. The curves update when the face changes, even if the number of edges changes.

Use **Datum** to create associative copies of datum planes and axes. These datums can be used for positioning features, as placement faces, as trim planes, or any of the other uses common to datums. You can control datum selection using the **Datum Type** list that displays in the dialog when you select **Datum**.

Select **Face** to extract solid faces or sheet bodies from another part. Sheet bodies are created in the work part. If you select more than one face or sheet, they are not sewn together automatically.

Selection allows you to filter based on **Selected Faces**, **Adjacent to Face**, or **All in Body**.

You can also enter a name in the **Name** text box and select **All Faces of Name** to select all faces with the specified name attribute.

Activate **Delete Holes** to remove any closed edge loops in the extracted face.

Use **Region** to extract a region of faces that act as a single feature during updates.

To create an extracted region of faces, use the two **Selection Steps** displayed in the dialog.

Select **Seed Face** to define a face within the extracted region.

Select **Boundary Faces** to define the outside or boundary of the desired region.

The seed face and all adjacent faces are included until reaching the boundary faces. You can preview the defined region by selecting **Preview Region**.

When you are working with a region where slots or pockets separate "island" faces from the seed face, you can activate **Traverse Interior Edges** to include these faces in the selection.

Activate **Delete All Openings** to remove all openings from the extracted work part face.

 Select **Body** to extract an entire solid body from another partand copy it into the work part. This is an excellent method of creating a parametric copy of a solid body within another part.

 Select **Mirror Body** to create a mirror image copy of a solid from another part.

Assemblies | WAVE Geometry Linker

Exercise: Creating WAVE geometry

1. Open **assy_7**.
2. Select **Assemblies | WAVE Geometry Linker**.
3. Click the icon representing the object that a linked copyis to be extracted.
4. Pick the corresponding object or objects from the displayed part.
5. If desired, activate **At Timestamp**.
6. If desired, activate **Blank Original**.
7. Click **OK**.

Summary:

In this chapter you learned how to:

- Understand and Apply the Assembly Structure Concept.
- Distinguish Between the Work Part and Displayed Part.
- Change the Work Part.
- Change the Displayed Part.
- View the Assembly Structure and Use the Assembly Navigator to Make Selections.
- Find Components and Perform Operations On Them.
- Create and Edit an Assembly Using the Bottom-Up Approach.
- Create and Manage Reference Sets.
- Display Only Necessary Geometry in Your Assembly by Replacing Reference Sets.
- Create a Top-Down Assembly.
- Use the Top-Down Approach to Design in Context of the Assembly.
- Create Sketches in the Assembly.
- Recognize the Importance of the Work Part.
- Remove Components.
- Transform, Rotate and Reposition Components.
- Understand the Mating Components Concept in an Assembly and Its Advantages.
- Create Mating Conditions and Constraints.
- Recognize Common Problems Encountered with Mating Conditions and Solve Them.
- Create an Explosion, Either Automatically or Manually.
- Show or Hide an Explosion from a View.

- Distinguish between Partially and Fully Loaded Parts.
- Use Load Options to Manage Large Assemblies.

Chapter 6

Freeform Modeling

After completing this chapter, you will be able to:

- Recognize the Differences between Conics, Circles, and Splines.
- Recognize the Differences between Splines with Varying Degrees.
- Identify the Continuity of a Curve.
- Create Bounded Planes.
- Create Ruled Surfaces.
- Create through Curve Bodies.
- Create through Mesh Bodies.
- Create Swept Bodies.
- Create Bodies Using Rows and Columns of Points.
- Create Bodies Using Points from a Data File.
- Create Bodies Using a Point Cloud.
- Understand the Common Options and Definitions for Sections.
- Create the Different Types of Sections. Create Different Types of Extension Sheets.
- Offset Sheet Bodies Using Constant, Variable, or Multiple Sheet Bodies.
- Quilt Surfaces Together Using Existing Sheet Bodies.
- Create Bridge Surfaces between Existing Sheets.
- Thicken Sheets to Create a Solid Body.
- Use Free Form Feature | Trimmed Sheet to Limit the Boundaries of Sheet Bodies.
- Create a Face Blend on a Sheet or Solid Body.
- Create a Soft Blend.
- Edit Parametric Free Form Features and Keep the Associativity.
- Extract Associative Copies of Faces.
- Extract Associative Copies of Face Regions.
- Extract Associative Copies of Bodies.
- Sew Sheet Bodies Together.
- Patch Solid or Sheet Bodies.

Goals of Design:

One of the main goals of a designer is to keep the design as simple as possible without compromising any design intent. At the modeling level, you accomplish this goal by using simple well-defined geometry from start to finish. To help meet this goal, Free Form Modeling involves a three-step process:

Create simple well-defined geometry that captures the design intent.

Analyze the sheet bodies/curves for continuity and flow as well astheir relationship to surrounding geometry.

Edit the sheet bodies/curves in the areas that need modifications.

Steps two and three may need to be repeated many times before the sheets meet all requirements.

You begin the Free Form Modeling process by applying this concept to the curves used to construct sheet bodies. Use this same process n to create the sheet bodies that will define the surfaces that make up the solid.

In the process of building a valid body, the math describing the solid becomes complex in a very short time. Because of this, it is important to keep the geometry defining the model as simple as possible from the conception of the design. All the elementary building blocks used in modeling points, lines, arcs, conics and splines are described by specific equations. Sheet bodies and solid bodies are defined by even more complex math. Feature operators such as trimming, sewing, hollowing, thickening and blending add more complexity to the final solid.

Simpler geometry will facilitate more success in creating a valid body, quicker computer response time, easier translation to other systems, smaller part file size, and ease in utilization for downstream manufacturing applications. You must have a good understanding of the applied math used in modeling to ensure that you fully describe the part in a concise and editable manner.

Unigraphics can create a variety of solids without using input curves. However, most Free Form Modeling applications will require the use of a previously created edge or generating curve. Because of this, it is important to control the complexity of these curves, as the complexity will be passed on to its associated body. The types of curves used to generate bodies are points, lines, arcs, conics, and splines. The choiceof curve used to model should always progress from simple to more complex. Using a simple curve minimizes the complexity of the resultant body. The following is a list of curve types in order of increasing complexity.

Advanced Curve Types:

There are three types of advanced curves: conics, splines, and law curves. We will discuss the first two.

Conics:

When the plane of intersection is parallel with the base of the cone defines a circle or arc. A Center point, Radius/Diameter, Start Angle and End Angle are components of Circle/Arc creation.

 Any plane of intersection that is not parallel to the cones centerline, base or half angle defines an ellipse. A Center point, Semi major Axis, Semi minor Axis, Start Angle, End Angle and Rotational Angle are components of an ellipse.

 When the plane of intersection is equal to the half angle of the cone defines a parabola. A Center point, Focal Length, Minimum Dy, Maximum Dy and Rotational Angle are components of a parabola.

 When the plane of intersection is parallel to the cones centerline defines a hyperbola. A Center point, Semi Transverse Axis, Semi Conjugate Axis, Minimum Dy, Maximum Dy and Rotational Axis are components of a hyperbola.

 A General Conic can be any of the four types: Circle/Arc, Ellipse, Parabola or Hyperbola. Unigraphics defines the type based on input data and the creation method chosen. Components of General Conic creation include Points, Slopes, Anchor, Rho and Coefficients. The seven General Conic creation methods use combinations of these components to define the Conic.

The Apex or Anchor is the intersection point of the Tangent Vectors that define theslope on each side of the Conic.

The Chord is the distance between the endpoints of the conic.

A Shoulder Point is any non-linear point between the end points of a conic used to define it and it is an interior point that lies on the conic curve.

The Rho value defines the elongation of the conic and what type it will be. It is the ratio of Y1 (the distance from the midpoint of the chord to the intersection of the conic with the chordal bisector) to Y2 (the distance from the chord midpoint to the apex).The value entered for Rho must be between .001 and .999.

When .001 < Rho < .500 an Ellipse is created.

When Rho = .500 a Parabola is created.

When .500 < Rho < .999 a Hyperbola is created.

Conics are defined mathematically as the result of an intersecton of a plane with a cone. The orientation of the plane relative to the cone type of conic created is defined by the. Unigraphics can create four types of conics: Circle, Ellipse, Parabola and Hyperbola. Each of these types has it's own feature operator. All four types can also be created using the General Conic feature.

Conics are defined by the intersection of a plane through a cone. The resultant conic type depends on the angle at which the cone has been cut.

Splines:

Splines created in Unigraphics are NURBS (Non-Uniform Rational B-Splines). They are created using defining data, poles or as the result of another operation. Defining data includes points, slopes and curvatures that describe the splines shape. Splines can also be generated as a result from operations such as intersect faces.

Spline Segments:

Splines can be created as either a single segment or multiple segment curve. A single segment spline is referred to as a bezier, while a multiple segment spline is a B-spline. A bezier spline is limited to 25 defining/creation points. When you need more than 25 points, you must create a multiple segment B-spline. A good first choice, which is also the simplest recommended spline, is a 3-degree bezier. If a 3-degree bezier cannot meet the curve requirements then a higher degree bezier can be used. When using a B-spline, use as few segments as are needed to describe the curve.

If a spline is a multiple segment, it is inherently stiffer than a single segment spline. Multiple segment curves can be manipulated to fit curvature and inflection with more accuracy than single segment curve. When editing a single segment spline, the modifications affect the entire spline. This is why single segment splines maintain a smoother shape. Editing a multiple segment spline affects only the segment containing the modifications and can create unwanted undulations.

Single segment splines are an excellent first choice to model with because of their simplicity. They should be used to generate sheet/solid bodies whenever possible. Typically, multiple segment splines are used to generate more complex sheet/solid bodies. Multiple segment splines are also the resultant splines that describe tangency, intersection, openings, and joint lines.

Spline Degree:

Every spline has a degree assigned to it. The degree of a spline refers to degree of the polynomial expression that defines it mathematically. The degree of the polynomial is defined by the largest exponential value contained within it. For example: **y = ax3 + bx2 + cx + d.** This equation has a degree of 3.

You can define and edit the degree for all splines. A bezier spline is always a single segment spline with a degree equal to the number of pole/points, minus one. Generally, the spline's number of poles determines the degree, minus the number of segments within that spline.

Most splines degree = number of poles - number of segments.

Bezier spline degree = number of points or poles - 1.

Splines can have a minimum degree of 1 and a maximum degree of 24. Most splines should be created as 3-degree single segment curves. If necessary, a 3 degree multiple segment spline can be used. In some situations, a 4 or 5-degree single segment spline is needed to define the tangency or curvature of the input curve. This is the recommended maximum degree for single segment splines. Operations such as projection or intersection create splines with segments of degree 3 by default.

The first two poles control slope and tangency at the beginning of a spline. The last two poles in the spline control slope and tangency at the end. The first three poles define curvature at the beginning of a spline. The last three poles in the splines define curvature at the end.

The following is a list of single-segment, low-degree polynomial equations and the associated curve.

A degree 1 curve has no tangency or curvature control at either end of the curve.

A degree 2 curve can have tangency control at either end. You can manipulate the tangency at either end but this may affect the tangency at the opposite end.

A degree 3 curve can have tangency control, independently, at both ends of the curve. You can manipulate the tangency at either end without affecting the opposite end.

A degree 4 curve can have tangency control, independently, at both ends. It also has curvature control at either end of the curve. The curvature can be manipulated at either end but it will affect the curvature at the opposite end.

A degree 5 curve can have tangency control, independently, at both ends. It also has curvature control, independently, at both ends. The curvature can be manipulated at either end without affecting the tangency or curvature at the opposite end.

Spline Stiffness:

Lower degree single segment splines are more flexible and modifications to the poles tend to change their overall shape. These splines reduce model complexity and enhance operation performance. Higher degree single segment splines are stiffer. The poles will have to be moved further to produce a significant change to the splines shape. Using higher degree splines also reduces the opportunity to transfer this data cleanly to other systems. When you use many points to define a spline, this may introduce unwanted oscillations.

It is possible to edit the stiffness of a spline. This modifies the shape of the spline by changing its degree while preserving the number of poles. The minimum degree allowed is 1 and the maximum degree is equal to one less than the number of poles. Increasing the degree makes the spline stiffer and less sensitive to modifications to its control polygon. Decreasing the degree reduces the stiffness of the spline, allowing it to follow the modifications made to its control polygon more closely.

Continuity:

The term continuity refers to the change in flow of a curve. Continuity applies in both curve creation and simplification. This requires you to manipulate the curve to ensure the desired type of continuity.

A curve can have C0, C1, and C2 continuity. These define the three curve definition types. The continuity types reflect the geometry combinations used in the creation of a curve. A brief description of each type is given below.

> **C0 continuity** is continuous with respect to position and contains a corner. A group of linear geometry joined by their end points defines this condition. With C0 continuity, the slope is non-continuous and there is no curvature.
>
> **C1 continuity** is continuous with respect to tangency and contains a tangent point. Linear segments joined by a circular tangent curve can define this condition. With C1 continuity, the slope is continuous but the curvature is non-continuous because of the linear sections.
>
> **C2 continuity** is continuous with respect to position, slope, and curvature. No linear sections exist in a C2 curve. Consequently, its shape is constantly changing.

 Splines are the most complex of the curve types. A Unigraphics spline is a curve that is not a line, arc, circle or conic. Splines may be three-dimensional whereas the other curve types are planar by definition. All splines created in Unigraphics are NURBS (Non-Uniform Rational B-splines). We use the term spline and B-spline interchangeably throughout this section.

Freeform Features:

You can create free form features as either sheet bodies or solid bodies. The type created depends on the modeling preference type and nature of the input curves. Unigraphics identifies sheet bodies and solid bodies as the same type of entity. They differ in that sheet bodies are a collection of faces and edges that define a boundary with zero thickness and no volume. Solid bodies have thickness and enclose a volume.

Most Free Form feature operators can create a solid if you meet the following conditions:

Preferences | **Modeling** body type is set to **Solid**.

The body is closed in both directions upon creation.

The body is closed in one direction and has planar ends.

Sheet bodies have many applications complimenting those associated with solid bodies. Sheets can mimic contours and shapes difficult to achieve using solid features. Use sheet bodies to trim a solid body and create contour on one or more faces. You can sew sheets together or use **thicken** sheets to create a solid.

When creating sheet and solid bodies, you must consider the complexity of the result. The complexity begins with the curves used to generate the sheet bodies. You create many curves just for the creation and manipulation of a body. Therefore, the body inherits the complexity found in the individual curves. The resulting body is not just a representation of a two or three-dimensional curve but a three dimensional entity with contour, area, and sometimes volume. Due to this fact, it is important to keep sheet bodies as simple as possible. Complex geometry continues to impact the model throughout the design process.

Faces are regions on the exterior of bodies enclosed by edges. Every face has a U and V vector direction assigned to it. The U vector refers to the row or section string direction, and the V vector refers to the column or sweep direction. The columns are roughly perpendicular to the rows.

In the creation of free form features, Unigraphics assigns isoperimetric curves to each face. The isoperimetric curves lie on the face in both the U and V directions. The relationship between the isoperimetric curves describes the face.

Unigraphics creates sheet bodies using **patches**. The system defines a patch internally to represent a portion of the sheet. The patches are pieced together to shapethe sheet body as it is defined mathematically. These patches do not create edges or seams in the surface and you can only view them through information functions. When creating sheet bodies, try to minimize the number of patches used to define it. Limiting the number of patches simplifies the sheet and improves downstream applications. Using more patches increases the localized control of curvature in the sheet but should be based on need.

You create sheet bodies as a single patch sheet or a multiple patch sheet. Some options allow you to choose the type and number you want to generate and other options use Unigraphics defaults.

A single patch creates a body consisting of only one patch referred to as "Bezier" sheets. A multiple patch creates a body consisting of a rectangular array of patches. When you edit a single patch sheet, the edit affects the entire sheet body shape. An edit to a multiple patch sheet body only affects the patches in the immediate area.

Sheet degree:

All sheet bodies are assigned a degree. Each sheet has a degree in both the U and V direction. The degree in either direction must be between 1 and 24. Lower degree sheets facilitate faster performance in subsequent operations and fewer problems in maintaining a workable model. Higher degree sheets may not translate to other systems as well, and when created through many points may produce unpredictable results.

Unigraphics sets the degree of some sheet bodies based on the type of sheet created and the complexity of the input curves. Others allow you to input the degree. You create most sheets with a degree of three. This is the default and the recommended start point.

A higher degree sheet is stiffer in the sense that you must move a pole a long way to produce any significant change in the overall shape of the body. Lower degree sheets are more pliable and modifications to the poles tend to alter their overall shape more.

Poles roughly correlate to the degrees of freedom of the sheet. To increase the degrees of freedom for a sheet, either increase the number of poles per patch (the degree) or increase the number of patches for the sheet.

The **Change Degree** option changes the degree of the sheet body by modifying the number of poles. This changes the degree while attempting to maintain the original sheet body shape and character. Increasing the degree of a sheet increases the degree of freedom. You can then alter the sheet in more directions and locations than before.

Stiffness:

Stiffness changes the degree of a sheet body as well. Increasing the degree creates a stiffer sheet by reducing the underlying patches and retaining the original pole count. This means that the overall shape of the sheet is less sensitive to modifications made to its poles. Decreasing the degree of a sheet reduces the stiffness by increasing the number of patches and retaining the original poles count. This allows it to mimic the undulations (reversal of curvature) of its control polygon more closely.

Continuity:

Continuity refers to the change in flow of sheet bodies as well as curves. Unigraphics uses C0, C1, and C2 continuity types to define sheet continuity, similar to curves. In sheets, continuity describes the manner in which faces meet along a common edge. The type of continuity contained in a sheet is dependent upon the types of geometry used in its construction. The type of continuity between sheets is dependent upon the transition from one sheet to the other across their common edge.

For sheets, C0 continuity implies two faces/sheets meet along a common edge.

C1 continuity implies two faces/sheets meet along a common edge and that the tangent plane, at each point along the edge, is equal for both faces/sheets. The transition across the edge is therefore tangent continuous.

C2 continuity implies two faces/sheets meet along a common edge, are tangent, and the rate of curvature change at each point along the edge is equal for both faces/sheets. The transition across the edge is therefore curvature continuous.

Smart Bodies and Associativity:

Smart Bodies remember their creation data (strings, tolerance, and parameters). You create most free form features as smart bodies except**Through Points**, **From Poles**, **From Point Cloud**, **Fillet** and **Foreign**.

Changing the geometry or expressions used to create the smart bodycauses the body to update to reflect the change. In addition, updating the smart body changes any child solid features. Because of this, it is important to keep the geometry simple and editable so updates cause no model errors.

Freeform Feature Sheets or Feature Sheets:

You create sheet bodies using **Insert | Free Form Feature**. You can also use **Extrude**, **Revolve**, and **Sweep along Guide** found on the **Insert | Form Feature** menu. Using these three operators and the following conditions cause the creation of sheets

Modeling Preferences set to **Sheet Body**.

- The use of an open section string.
- The use of a non-planar section string.

There are many commands in the **Insert | Free Form Feature** menu that create sheet bodies. Some can be set to create solid bodies as well as sheets. It becomes a question of which type of sheet (Free Form Feature or Feature) you need. We list some of the comparisons and limitations below to help in the decision.

Sheet bodies created in both applications have associativity to the section strings or sketches used to create them. You can analyze and edit the parameters of each type of sheet. You can use both types of sheet bodies in later operations to build and refine the solid model. There are, however, a few differences between them.

Feature sheets retain the sharp corner condition of the section string while you must set the tolerance to **0.0** on Free Form sheets to maintain a sharp corner. If the tolerance is not **0**, Unigraphics incorporates a small blend into the corner.

Feature sheets only allow the use of a single section string while free form sheets allow the selection of multiple section strings. This is a major advantage when you require more control over the contour of the sheet.

Feature sheets have minimal control over the sweep of the section string. **Extruded Body** is a linear transition, **Body of Revolution** is a sweep about an axis, and **Sweep Along a Guide** is a transition from the start point of a guide to the end point. Free Form sheets, on the other hand, allow for many orientationmethods of the section string. They also allow multiple guide strings and the use of spines. Some allow for the definition of the transition (Linear or Cubic). Using cross strings that define the sheet in the direction opposite to the section strings is available. Free Form sheets contain the option of using points to define the sheet or orientation of the section strings.

When creating simple sheet bodies requiring minimal contour and control of definition the feature sheets are a good choice. If you need a more complex sheet, use free form sheets. Free form sheets have many different methods of creation and also have many more input choices for each method. Free form sheets allow control of the contour over every portion of the resultant body and provide more flexibility with editing than feature sheets.

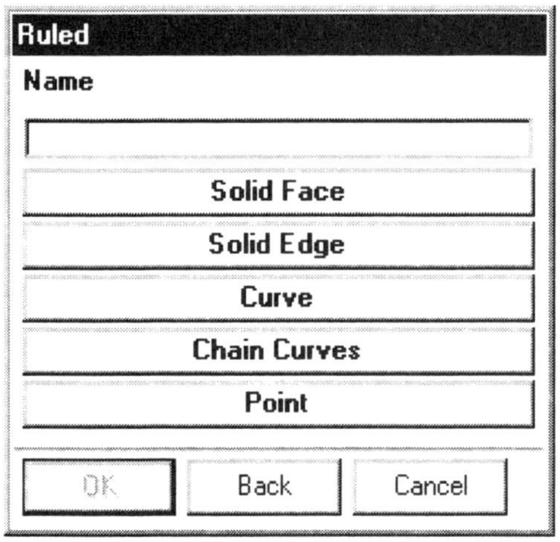

Section String Selection:

Any free form features requiring curves or edges to define their shape use the **Section String** dialog to select the geometry. Although the dialog names change depending on which command you use, the options within that dialog remain the same.

Solid Face lets you pick a face as a section string segment. The edges of the face define the section string segment.

Solid Edge lets you pick individual edges of a solid to define the section string segment.

Curve lets you pick curves.

Chain Curves lets you pick a contiguous string of curves. After picking the last section string, select **OK** to finish the current string and then select **OK** a second time to finish picking section strings.

Point shows only in the Ruled and Through Curve Mesh dialogs. Use this when you create a triangular sheet body. You can only pick an endpoint of a curve.

Notes: Select **Chain Curves** when using curves to define a section string because any gaps in the string causes the chaining to fail. This method is a good check to ensure a closed profile.

If you pick one sketch curve, all the curves in the sketch highlight unless you use the **Curve** or **Chain Curves** options.

Creating Bounded Planes:

**Insert | Form Feature | Bounded Planes**

Create a planar sheet body quickly. You construct the bounded plane sheet body by defining a planar profile. Curves, solid faces, solid edges, or any combination of the three may define the profile. All geometry selected must define aregion on a plane or an error message appears.

Exercise: Creating a bounded plane surface

1. Open **ff_1**.
2. Select **Insert | Form Feature | Bounded Plane**.
3. Pick the appropriate faces, edges, and curves to enclose an area on a plane
4. Select **OK**.

Alignment Methods:

Alignment methods define how each section string is broken up to define the transition to the other. Each of the following alignment methods, **Parameter**, **Arc Length**, **By Points**, **Distances**, **Angles**, and **Spine Curve**, uses a set of internal reference points on each of the section strings. A line created through each corresponding point is referred to as an isoperimetric curve. The isoperimetric curves define the way the shape of the first section string is communicated to the second.

Various free form features use alignment methods. Any of the free form features using section strings use alignment methods to control the shape of the bodies created.

Parameter spaces the points through which the isoperimetric curves pass at equal parametric intervals along the section strings. The parametric increments on the section strings are calculated by defining each string as an equation with one variable. The X and Y point values on the section strings the isoperimetric curves pass through are then calculated in terms of this variable. Each increment in this variable creates a corresponding point value on the section string. Similar points on each section string are then used to define the isoperimetric curves. The entire length of the section string is used in parameter alignment.

Arc Length spaces the points at equal length intervals along the entire length of the section strings. Similar points on each section string are then used to define the isoperimetric curves.

By Points allows you to assign the points through which the isoperimetric curves pass on each section string. You can align points between section strings with very dissimilar profiles and control the definition of the resultant sheet body. Begin by picking an alignment point on Section String One, and then pick a corresponding alignment point on Section String Two. Continue to choose alignment point pairs until enough definition exists to create the desired sheet. The beginning and endpoints defining the section strings cannot be selected as alignment points. Unigraphics automatically selects them as initial alignment points. It is possible to use a defining point in more than one point pair. Each section string must be made up of more than one curve or edge. Only the endpoints of each curve/edge segment in a section string can be chosen as alignment points. This method uses the entire length of the sections.

By Points allows you to create sharp edges and faces that terminate at a point.

Distance spaces the points through which the isoperimetric curves pass at equal distances normal to a specified vector direction. Unigraphics constructs planes normal to the specified vector direction and intersects them with the section strings. The section string that fails to intersect these normal planes first limits the resultant sheet.

The section strings must be able to be subdivided by planes perpendicular to the direction vector, otherwise the Distance error message appears.

Angle spaces points defined by the intersection of the section strings with radial planes about a specified axis to define the isoperimetric curves. The section string that fails to intersect these radial planes first limits the resultant sheet.

Spine Curve creates planes perpendicular to a selected curve. These planes are then intersected with the section strings to generate the points for the isoperimetric curves. This option is similar to the **Distance** option except the spine curve can have curvature.

Notes: The **Parameter** option is a good choice when the section strings are similar geometrically.

Use the **Parameter** option when you do not want to approximate the section strings.

Use the **By Points** option when sweeping section strings that have different shapes and a specific translation is needed.

Creating a Ruled body:

 **Insert | Free Form Feature | Ruled**

Section strings are on opposing ends of the sheet body. The contour of the section strings and the linear transition between the section strings define the sheet body. The section string can be composed of any combination of solid faces, solid edges, curves, or a point. The resultant sheet body is associative to the section strings. Any changes in the section strings reflect in the ruled sheet.

Tolerance specifies the maximum distance a sheet can vary from the section string. A tolerance value of 0.0 creates a sheet that matches the curve exactly. Decreasing the tolerance value increases the amount of defining data the system stoes for the resultant sheet body. Increasing the tolerance value allows the system to approximate the section string and decrease the complexity of the sheet body.

Temporary Grid display allows the viewing of the U and V isoperimetric curve flow used to build the ruled sheet. Inputting values into **Temp u Count** and **Temp v Count** create the temporary grid. If the grid is not satisfactory, use another alignment method to create the sheet. The grid displays on the sheet until you accept the surface. This is onlya representation of the isoperimetric curves but allows a visualization check and an opportunity to change alignment methods before final creation. The U direction represents the section strings and the V direction represents the isoperimetric rulings connecting the section strings. A larger ruled sheet more complex may require more temporary U and V curves to determine the quality of the sheet to be created. Making the value **0** for both settings causes the system to skip the temporary grid and confirmation.

Exercise: Creating a Ruled Surface

1. Open **ff_2**.
2. Select **Insert | Free Form Feature | Ruled**.
3. Pick **Section String 1** and continue to click **OK** until the Cue line prompts you for Section String 2 and the **Ruled** dialog for picking the second section string.
4. Pick **Section String 2**.
5. Click **OK**.
6. Select an **Alignment** setting.
7. Type in the appropriate values for the **Tolerance**, **Temp u Count** and **Temp v Count**.
8. Click **OK**. If you entered values in the **Temp u Count** and **Temp v Count** a dialog appears to accept the surface.
9. Click **Yes** to create the ruled surface.

Note: Why did the surface have a twist in it? You picked your section strings on opposite ends and your vectors were pointing in opposite directions.

Remember to read the Cue line. It is easy to pick the first and secondsection string while the prompt is still asking for the first section string. An easy way to determine you have finished done specifying a string is by the appearance of a vector on the section string.

To ensure sharp corners are retained use a tolerancevalue of **0.0**. If any value other than **0** is used in the **Tolerance** text box, an approximation of the curves is used to create the sheet body.

Creating through Curve Bodies:

 **Insert | Free Form Feature | Through Curves**

It is very similar to the **Through Points** option but instead of specifying rows defined by points, specify rows defined by section strings. The amount of sectons can vary between 2 and 150. The resultant sheet body from the Through Curve method is associative to the section strings. Any changes in the section strings are reflected in the sheet body.

Patch Type specifies whether you create a Single or Multiple patch sheet body. You must select at least two strings when using the **Single** option (up to a maximum of 25). **Multiple,** allows for a transition with more curvature between section strings.

With **Closed in V**, the sheet closes along the columns (the V direction). If you select closed sections strings, Unigraphics creates a solid body.

V Degree refers to the degree of the polynomial defining the sheet body in the "V" direction (the direction of the column). Use a degree of three, which is the default, for better downstream application. This option is only available when you set the **Patch Type** to **Multiple**.

For single patch sheets, the degree in the V direction is always one less than the number of section strings used. You determine the degree in the U directionby what you used for the tolerance. **Tolerance** defines the maximum distance between the section string and the sheet body. If a value of **0.0** is used, the curve with the highest degree determines the degree of the U direction of the resultant body. With a tolerance greater than **0.0**, the U degree is 3.

First and Last String:

> **No Constraint** does not assign any constraints to the sheet; the section string geometry determines the curvature and slope of the sheet body.
>
> **Tangency** lets you assign a tangency constraint to either the first, last, or both section strings. This constraint uses contiguous faces from another sheet body or solid to determine the tangency.
>
> Unigraphics projects the section string onto the constraint face and makes the slope of the new sheet body tangent. If the section string cannot project onto the constraint face, you receive an error message. Unigraphics creates a smart sheet relationship between the new sheet body and the constraint face. If you modify the constraint face, the new sheet body automatically updates. If you delete the constraint face, the through curves sheet becomes unparameterized.

Curvature is similar to **Tangency** but instead of making the slopes equal, it makes the curvature the same. Both **Tangency** and **Curvature** produce a tangent sheet. The Curvature constraint maintains the curvature of the constraining face for approximately 10 percent of the new sheets overall length. When you edit the sheet, you can create the curvature constraints.

When you set **First Section String** or **Last Section String** to either **Tangency** or **Curvature**, the **Direction** list becomes active. There are three settings to choose from to determine the shape of the cross section for the Through Curve feature.

Not Specified matches tangency or curvature to the constraining faces. The cross section of the through curve body does not lie at a specific location. Unigraphics approximates the cross section between the constraining surfaces.

Isoperimetric produces a resulting body matching tangency or curvature to the constraining faces. The cross section of the resulting feature follows the isoperimetric curves of each constraining face. **Isoperimetric** produces the most accurate constraint condition but, based on the complexity of the constraining faces, can be unpredictable.

Normal matches tangency or curvature at the constraining faces. The cross section of the resulting sheet is normal to the constraining faces at the section strings.

Exercise: Creating a Through Curve Body

1. Open **ff_2**.
2. Select **Insert | Free Form Feature | Through Curves**.
3. Pick Section String #1 in the Graphics window
4. Click **OK**, then continue to pick the desired sections strings, making sure to click **OK** after each section string pick.
5. Click **OK** twice after picking your last section string.
6. Set the **Patch Type**.
7. Set the **Alignment**.
8. Verify the **V Degree** and activate **Closed** if desired.
9. Type a value in the **Tolerance** text box.
10. Select the appropriate constraint for the first and last section string.
11. Click **OK**. When using constraint methods, you must specify the constraint faces.

Notes: You can pick the edges of a sheet body to define a section string segment.

It is easy to lose your place when creating a free form feature through curves. You must remember to tell Unigraphics you have completed the definition of one string and you are beginning the definition of another. Pay close attention to the Cue line and Status line.

The Status line prompts you to "Select OK to Finish Selecting String #." Select **OK** to start the next section string. After each string is specified, a vector displays. This is a good identifier to let you know you are now going to pick the next section string.

When selecting the section string, be consistent with the direction of the vectors corresponding to the strings. If you are not consistent and the vectors oppose each other, Unigraphics creates the sheet with a twist.

Create B-splines with a low degree. High-degree B-splines have many points along the section curve, unnecessarily complicating the resultant sheet.

If you want the sheet to match the section strings exactly, give **Tolerance** a value of **0**. We also recommend retaining sharp corners from section strings.

Through Curve Mesh Bodies:

Through Curve Mesh creates a sheet body by defining primary and cross strings. The section strings in the U direction are the primary strings and the section strings in the V direction are the cross strings. This option provides more control of the shape between the section strings than the **Through Curves** option. These curves create a mesh of curves that defines the sheet.

Section Strings:

As with the **Through Curves** option, you can pick faces, edges, curves and sketch curves to define the primary and cross strings. It is important that you are consistent when picking both sets of section strings. The set of section strings selected as cross strings is roughly perpendicular to the primary curves. The minimum number of primary strings is two and the maximum is 150. Pick the strings in an orderly manner, moving from one side to the other. When creating a through-curve mesh body, primary strings cannot have the same endpoint. Cross strings can share the same endpoint if you create a triangular sheet body.

If you need an irregularly shaped surface, you can pick a point as the first and/or last primary string. When using a point, be sure to pick it during the primary curve selection because it is not available during cross string selection.

The use of a spine string is optional. Using a spine string helps control the orientation of the cross strings as the body is created.

The spine for a through curve mesh feature must run in approximately the same direction as the cross strings. Unigraphics defines planes normal to the spine. These planes define the flow of the V direction isoperimetric curves by using their intersection with the cross curves.

Creating through Curve Mesh Bodies:

Insert | Free Form Feature | Through Curve Mesh

Create a 3-degree, multiple patch sheet body that is associative to the primary and cross strings. Changes in the strings are reflected in the sheet body.

Emphasis selects the set of curves you want the body to favor in case the U and V section strings do not intersect. You use the options **Primary** or **Cross**-when the primary curves do not intersect the cross curves.

> **Both** places equal emphasis on the primary strings and the cross strings. If the two sets of strings do not intersect but lie within the specified intersection tolerance, Unigraphics creates the sheet body lying halfway between the curves.
>
> **Primary** places the emphasis on the primary strings. Unigraphics uses the cross strings for the shape in the cross direction but the sheet body might not pass through the cross strings.
>
> **Cross** places the accentuation on the cross strings. Unigraphics uses the primary strings for the shape in the primary direction but the sheet body might not pass through the primary strings.

Intersection Tolerance:

This allows you to specify a tolerance between the primary and cross curves. This tolerance is the maximum allowable distance that the primary curves can be from the cross curves and must be greater than zero. If the curves do not intersect within the tolerance, an error message appears and the incorrect string pairs highlight. The error message appears for each pair of curves out of tolerance.

First, Last and Cross Strings:

> **No Constraint** is the system default. This option does not assign any constraints to the sheet. Curvature and slope of the sheet body are determined by the geometry used to create the sheet.
>
> **Tangency** let's you assign a tangency constraint to either the first, last, or both section strings. This constraint uses a face from another sheet body or solid to determine the tangency. The process is similar to the **Through Curves** option except you can assign up to four tangency conditions. Unigraphics projects the section string onto the constraint face and makes the slope of the new sheet body tangent. If Unigraphics cannot project the section string onto the constraint face, you receive an error message. Unigraphics creates a smart sheet relationship between the new sheet body

and the constraint face. If you modify the constraint face, the new sheet body automatically updates. If you delete the constraint face, the through curve mesh sheet becomes unparameterized.

Curvature **is similar to the option found in** Through Curves **except you can assign up to four curvature conditions. This option is similar to** Tangency **but instead of making the slopes equal, it makes the curvature the same. Both** Tangency **and** Curvature **produce a tangent sheet. The** Curvature **constraint maintains the curvature of the constraining face for approximately 10% of the new sheets overall length. Curvature constraints can be created with the sheet creation or added later by editing the sheet.**
Construction options:

These determine the complexity of the surface.

Normal creates the most complex surface.

Spline Points uses the points and tangent values at the points on the section strings. Unigraphics creates the new sheet through these points and not necessarily the curves. It is important to note that the section strings must be single B-curves with the same number of defining points.

Simple creates a surface using the most complex curve to determine the surface complexity.

Exercise: Creating a through curve mesh surface

1. Open **ff_4**.
2. Select **Insert | Free Form Feature | Through Curve Mesh**.
3. Pick the first primary curve in the Graphics window. This can be made up of a series of continuous curves.
4. Click **OK** until the direction arrow appears.
5. Pick the next primary curve. Repeat this process until you pick all the primary curves.
6. After picking the last primary curve, click **OK** twice.
7. Pick the first cross curve in the Graphics window.
8. Click **OK** and pick the next cross curve. Repeat this step until you pick all cross curves.
9. After picking the last cross curve, click **OK** twice.
10. Pick a Spine String if needed. Click **OK**.
11. Specify the **Emphasis**, **Intersection Tolerance**, and the **Construction Options**.
12. Select the appropriate constraint for the First and Last Primary and Cross Strings.
13. Click **OK**.

Simple Construction Method:

By using the **Simple** option, you can create a surface with a minimum number of poles and patches. The degree, poles and patches of a simple surface relate directly to the degree, poles and knots of the defining splines. This option is available in the **Through Curve Mesh** and **Through Curves** dialogs.

The **Simple** method minimizes the complexity of the surface by keeping the poles and patches to a minimum. When you edit these surfaces, update time decreases due to the simpler surface. **Simple** also keeps the part file size smaller.

For a through-curve mesh surface, the highest degree curve in the primary and cross directions determines the simple surface's U and V degrees. The curve with the highest number of poles determines the surface poles in the U and V direction. The curve with the highest number of segments determines the number of patches in the U and V direction of a through-curve mesh surface.

For a through-curves surface the curves only affect the number of poles in the U direction and the number of segments in the U direction. The highest degree curve determines the surface degrees in the U direction only. You still must input the number of degrees for the V direction.

In most cases, the deviation between surfaces using **Simple** compared to **Normal** is insignificant. The Information windows show two surfaces created using the same curves but different construction methods. Notice the difference in the number of poles, seams and patches.

Note: Use all 3-degree curves to further minimize the surface complexity.

Freeform Swept Features:

Swept features create sheet or solid bodies using a profile(s) swept along one, two, or three guide strings. This creation method is similar to the **Insert | Form Feature | Sweep Along Guide**. However, you have more options to help define the sheet as the section string(s) sweeps along the guide string. You must specify at least one section string, with a maximum of 400, and at least one guide string with a maximum of three. The section string(s) defines a cross section of the resulting body, while the guide string(s) defines the path of the sweep.

Specifying one guide string:

When you use one guide string, you have options to control orientation and scale of your resultant body. You can control the shape using other geometry, specifying certain parameters, or even by using expressions. It becomes somewhat more difficult to determine the shape using these methods.

Specifying Multiple Guide Strings:

When using multiple guide strings, there is no need, nor will you have the ability, to assign the orientation and scaling methods available with a single guide string feature. The additional guide strings determine the orientation and scale. There can be a maximum of three guide strings. Unigraphics breaks each guide string into a series of points by either **Parameters** or **Arc length**. Unigraphics uses these points to orient the section string. **By Points** is also available for multiple section string features.

When you create a swept feature using two guide strings, the first string acts as the guide and the second determines the orientation. You determine the scale using the scaling method dialog. The two available scaling methods are **Laterally** and **Uniformly**. These options display in the **Scale** dialog after you select all guide and section strings.

When you create a swept feature using three guide strings, the three guide strings determine the path, orientation, and scale. The corresponding points of each section string determine the orientation.

Creating Swept Bodies Using One Guide String:

 **Insert | Free Form Feature | Swept**

Interpolation Method:

If you select two or more section strings, you must specify the transition between the section strings. The **Linear** or **Cubic** options in the **Interpolation Method** dialog define this transition.

> **Linear** interpolation defines a linear rate of change between section strings to create the body. This yields a straight-line effect in the transition area.
>
> **Cubic** interpolation defines a cubic rate of change between section strings. This yields an s-shaped effect in the transition area.

Tolerance controls how closely the swept feature follows the section strings. If you want the resulting body to conform exactly to the section string, type **0.0** for the **Tolerance**. This transfers the sharp corners of the section strings into the resulting body.

Orientation controls the relationship of the section string to the guide string. A series of local coordinate systems created at points along the guide string define the orientation of the section string as it sweeps along a guide string. You cannot edit the orientation once you create the body. These local coordinate systems are defined in the following manner. The first axis (Z) is the tangent vector to the guide string at that point. Choosing one of the orientation options creates the second axis (X). The third axis (Y) is a normal to the other two.

Fixed orients the local coordinate system so the X- Y plane is normal to the tangent vector. The nature of the guide string defines the direction of the X and Y-axis.

Face Normals orients the local coordinate system so the second (X) axis is parallel to the normal of the face selected for orientation at each point. If the desired result is the orientation of the local coordinate systems shown in the figure, specify the face normal to the X- Y plane shown at each point.

Vector Direction orients the local coordinate system so the second (X) axis is parallel to a vector defined using the Vector Subfunction. If the desired result is the orientation shown in the figure, the vector direction specified must be the Absolute Coordinate System Z direction.

If Unigraphics cannot create a coordinate system for the section, an error message appears. This may happen if the vector is parallel to the Z-axis of the local coordinate system at some point along the guide. Two parallel vectors cannot define a local coordinate system.

Another Curve bases the local coordinate systems on a set of isoperimetric curves created between the guide string and the orientation curve. These isoperimetric curves are the same as those created using the Ruled Sheet feature on the same input strings. The second (X) axis aligns with these isoperimetric curves at each point.

If the orientation string and the guide string are not coplanar, the swept feature rotates as it follows the guide. Using the **Another Curve** method allows the selection of a spine string. The **Spine** option is the same as the Spine method discussed in the Ruled section. This determines the second axis of the local coordinate system.

A Point uses isoperimetric curves created between the guide string and a chosen point to align the second (X) axis. These isoperimetric curves are the same as those for a similarly created ruled sheet. One end of the section stays in a fixed position (aligned with the chosen point) while the other end sweeps along the guide string.

Angular Law bases the angular rotation of the section on the **Law Definition** options.

Forced Direction fixes the orientation of a section string plane normal to a vector.

When you select this, the dialog for defining the orientation vector displays. **Datum Axis** allows you specify an associative vector.

After you specify the direction, sweeping the section along the guide using planes parallel to the specified direction creates the feature.

Scaling allows the section string to increase and/or decrease in size as it travels along the guide string. The **Swept** dialog only appears when you pick one guide string.

Constant will prompt you to type in a scale factor for the section string. The default value is **1.0**. If you input a different value, Unigraphics scales the section string using the start point of the guide string as the scaling point. This is a uniform scale of the section string along the entire length of the guide.

Blending Function allows you to input a start scale at the first section string and an end scale at the second section string. It also lets you select either a **Linear** or **Cubic** transition from string one to string two using the **Interpolation Method** dialog.

Once you choose an Interpolation Method, you enter the start and end values in the **Enter start and end scales** dialog.

If you enter a scale value other than **1** for the start or end values, the interpolation specified with the **Blending Function** overrides the previously specified **Interpolation Method**.

Another Curve uses a curve or set of curves to determine the scale of the section string. The scale of the section as it is swept is directly proportional to the distance between the guide string and a specified curve or solid edge.

A Point uses a fixed point to determine the scale of the section segment. This option works on the same principle as the **Another Curve** option except the distance is measured along the guide string to the point. Use this option when creating a three-sided swept feature, where you also use the scaling point for the orientation point.

Area Law uses the Law definition dialog to determine the actual cross sectional area of a resulting solid body. Normal planes to the guide string determine the cross section cuts. You must use closed section strings for this option to function.

Perimeter Law is similar to the **Area Law** option except that the Perimeter Law Definition dialog controls the cross sectional perimeter instead of controlling the area.

Exercise: Creating a Swept Feature Using One Guide String and One Section String

1. Open **ff_5**.
2. Select **Insert | Free Form Feature | Swept**.
3. Pick the first guide string.
4. Click **OK** twice.
5. Pick the first section string.
6. Click **OK** twice.
7. Select the desired **Alignment Method** setting and enter a **Tolerance**.
8. Click **OK**.
9. Click an orientation setting.
10. Click a scale setting and define the scale method.

Exercise: Creating a Swept Feature Using One Guide String and Multiple Section Strings

1. Open **ff_6**.
2. Select **Insert | Free Form Feature | Swept**.
3. Pick the first guide string.
4. Click **OK** twice.
5. Pick the first section string.
6. Click **OK**. Keep picking sections strings until you have specified all you need.
7. Click **OK** twice after the last section string.
8. Click either **Linear** or **Cubic**.
9. Select the desired **Alignment Method** setting and enter a **Tolerance** value.
10. Click **OK**.
11. Click an orientation setting.
12. Click a scale option and define the scale method.

Creating Sheet Bodies Using Points and Poles:

You can create sheet bodies using points where the sheet body passes through the points or the points "guide" the shape of the sheet body. You can also create a sheet body using an irregular grouping of points.

Sheet Bodies through Points:

This creates a sheet body defined by an array of points. The points make up rows and columns, which define the U and V contour. The point set defines the form in a rectangular grid with groups of points in the row and column (U and V) directions.

Corresponding points in each row define a column. When you use the same number of points in each row, a much cleaner and more predictable sheet body results. After you define the points, Unigraphics creates a Through Points sheet body that passes through all points within the specified modeling tolerance. The Through Points method always results in an unparameterized sheet body.

Sheet Bodies through Poles:

This creates a sheet body using defined points as poles or control vertices for the face. The difference between a From Poles and a Through Points sheet body is that the points you pick are poles that control the size and shape of the body. This option usually creates a smoother sheet than one created using through points. The **From Poles** option always results in an unparameterized body. You can pick points using the Point Constructor dialog or using Points From File.

How sheet bodies through points and through poles differ:

Through Points guarantees that the sheet passes through the points. However if you do not place the points perfectly unwanted flat spots or points of inflection can occur.

From Poles provides control over the overall shape of the sheet body but the sheet body does not necessarily pass through any particular pole. The sheet body always passes through the start and end rows of points.

Sheet Bodies from a Point Cloud:

This creates a sheet body quickly and without much manipulation of data points. It uses random data point locations and does not require points to be in rows and columns. This is an extremely useful feature for digitized data. Unigraphics best fits a sheet body through all the points selected. This operation is similar to the Fit Spline concept. The resulting sheet is much smoother than one created from the same points using the **Through Points** method, but does not conform as closely to the defining points. **From Point Cloud** always creates an unparameterized sheet body.

Note: If you need a non-analytical aesthetic sheet body, use the **From Poles** method to create a smoother surface.

Through Points:

 **Insert | Free Form Feature | Through Points**

Create a sheet body defined by a rectangular grid of points that define the U and V directions of the sheet body.

Use **Patch type** to create a sheet body consisting of **Single** or **Multiple** patches. If you select **Single**, the number of points in a row determines the U direction degrees and the number of rows specified minus one determines the V direction degrees. If you select **Multiple**, you determine the degrees using **Row** and **Column Degrees**.

Closed along works in conjunction with the Multiple Patch type. If the patch type is set to **Single**, this option is not available. There are four **Closed Along** options.

Neither creates a sheet body through the points selected.

Row connects the last point in each row with the first point in that row. If the end columns are planar, Unigraphics creates a solid body.

Column connects the last point in each column with the first point in that column. If the end rows are planar, Unigraphics creates a solid body.

Both close the body in both U and V directions (rows and columns). This results in a solid body.

Row Degree:

If you set the **Patch Type** to **Multiple**, **Row Degree** becomes active. This option controls the degree of the polynomial that defines the sheet body in the U direction (in the direction of the row). You can use a degree between 1 through 24 but we recommend the default degree of 3.

Column Degree:

If you set the **Patch Type** to **Multiple**, **Column Degree** becomes active. This option controls the degree of the polynomial that defines the sheet body in the V direction (in the direction of the column). You can use a degree between 1 through 24 but we recommend the default degree of 3.
As an example, a **Row Degree** of 3 and a **Column Degree** of 3 requires you to select a minimum of four points in a row and four rows of points. You must pick at least one more point in the row or column than the degree number.

Points from file will allow you to use a point data file to specify points used in the creation of a sheet body. When you select this, Unigraphics prompts you to select an existing data file. You can make the points relative to the Absolute Coordinate System or the WCS. After you select the data file, Unigraphics automatically enters the data points and creates the sheet.
When creating a point data file, input the data as in the format shown below.

ROW 1

1.25 2.125 3.00 - represents point 1 of row 1
1.125 2.065 3.00 - represents point 2 of row 1
1.00 2.000 3.25 - represents point 3 of row 1
0.875 1.975 3.375 - represents point 4 of row 1

ROW 2

1.25 2.125 4.00
1.125 2.065 4.00
1.00 2.000 4.25
0.875 1.975 4.375

The rows are separated by the header "ROW." Unigraphics is looking for ROW and ignores anything that comes after the word "ROW." The three inputs are always X, Y, and Z coordinates and can be spaced by either tabs or by spaces.

Point Selection Options:

Chain From All allows you to pick an existing point object as the start point and another existing point as the endpoint. Unigraphics picks all points that lie between the start point and end point. Use this option only when the distance between rows is greater than the distance between the points defining the row.

Use **Chain Within Rectangle** to drag a rectangular boundary around the points in the Graphics window. Unigraphics prompts you to pick a start point and end point to determine the proper order of the points within the rectangle.

Chain Within Polygon works exactly like **Chain Within Rectangle** except you pick the points with a polygon instead. After the polygon is defined, pick a start point and end point to determine the proper order of the points.

You can use **Point Constructor** to pick any type of point.

Exercise: Creating a through Points Sheet Body

1. Open **ff_7**.
2. Select **Insert | Free Form Feature | Through Points**.
3. Select your desired options and click **OK** to display the **B Surface Points, Row 1** dialog.
4. Click a **point specifying option**.
5. Pick the first row of points and click **OK** when complete to display the **Point Specification** dialog.
6. Click **Yes** to redisplay a point selection dialog. Clicking **No** returns you to **Point Specification** dialog.
7. Pick the remaining rows of points until the **B Surface** dialog displays.
8. Click **Specify Another Row** to return to the **Point Specification** dialog. If you are done specifying rows, click **All Points Specified** to create the sheet body.

From Poles:

 **Insert | Free Form Feature | From Poles**

This creates a sheet body using defined points as poles or control vertices for the face. The options within the **From Poles** dialog function the same as in the **Through Points** dialog.

Exercise: Creating a sheet body using from poles

1. Open **ff_8**.
2. Select **Insert | Free Form Feature | From Poles**.
3. Select your desired options and click **OK**.
4. Pick the first row of points and click **OK** when complete to display the **Point Specification** dialog.
5. Click **Yes** to redisplay the **Point Constructor** dialog.
6. Pick the remaining rows of points until the **B Surface** dialog displays.
7. Click **Specify Another Row** to return to the **Point Constructor** dialog. If you are done specifying rows, click **All Points Specified** to create the sheet body.

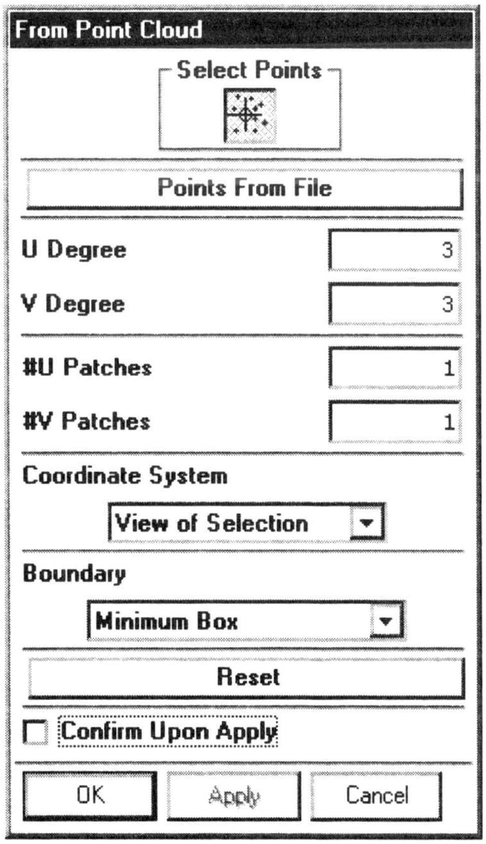

From Point Cloud:

Insert | Free Form Features | From Point Cloud

With this you can create a sheet body quickly using random data point locations not requiring points to be in rows and columns.

Use this to pick points for the definition of the sheet body. You can pick points individually or drag a rectangle around a group of points. When the points are selected, you can deselect the points by pressing the SHIFT key as you pick the point to deselect.

U and V degree controls the degree of the sheet body in the U and V directions. The default degree is 3 but you can change it to any degree between 1 and 24. We recommend a degree of 3.

#U and #V patches controls the number of patches in both directions. The combination of degree and patches controls the curvature of the surface. The degree and number of patches determine the number of points required to create the sheet body. Adding the U degree and U patches and multiplying this by the sum of the V degree and V patches define the number of points required. If there are not enough points for the specified degree and patches, an error message appears.

Coordinate system uses the coordinate system you specify to determine the U, V, and sheet normals. The X- and Y-axis determines the U and V directions of the sheet body respectively, and the Z-axis determines the normal. The X- and Y-plane of the coordinate system should be as parallel to the sheet as possible. The important thing to remember is that when you view the points looking down the Z-axis, the points do not create a surface that overlaps itself. The five options for the Coordinate System are:

> **View of Selection** creates a U-V grid that is the normal to the plane of the screen. This creates a U vector that points to the right and a V vector that points up. The normal vector points out of the screen. After you define your points, you can rotate the view without affecting your view of selection direction. If you pick any more points from after rotating your view, the view of selection direction changes to your new view.
>
> **WCS** uses the current WCS as the coordinate system
>
> **Current View** uses the coordinate system of the current work view.
>
> **Specified CSYS** selects the coordinate system previously defined by the **Specify New CSYS** option.
>
> **Specify New CSYS** allows you to define the coordinate system using the Specify Coordinate System dialog.

Boundary defines the bounds or edges of the sheet. From Point Cloud creates a rectangular sheet body that is projected to the U-V plane. Unigraphics always creates a rectangular sheet body regardless of the shape of the points selected. There are three options for boundary definition.

Minimum Box creates a minimum rectangle on the U-V plane enclosing all the points specified.

Specified Boundary uses the boundary defined by selecting the **New Boundary** option.

New Boundary defines the boundary using the **Point Constructor** dialog. Unigraphics only allows you to pick four points that define the corners of the rectangle.

Reset deselects all points and resets the settings to the default settings.

Exercise: Creating a Sheet body with From Point Cloud

1. Open **ff_9**.
2. Select **Insert | Free Form Features | From Point Cloud**.
3. Click **OK** in the Error message to display the **From Point Cloud** dialog.
4. Pick the points in the Graphics window.
5. Enter the **U** and **V Degree** of the surface.
6. Specify the **#U** and **#V Patches**.
7. Set the Coordinate System from the list.
8. Specify the Boundary of the new sheet from the list.
9. Click **OK** to display the Fit Information message.
10. Click **OK** from the message to create the sheet body.

Note: The area where the surface normals are parallel to the Z-axis produce the best surface. The closer the surface normals are to the Z-axis, the more accurate the results are.

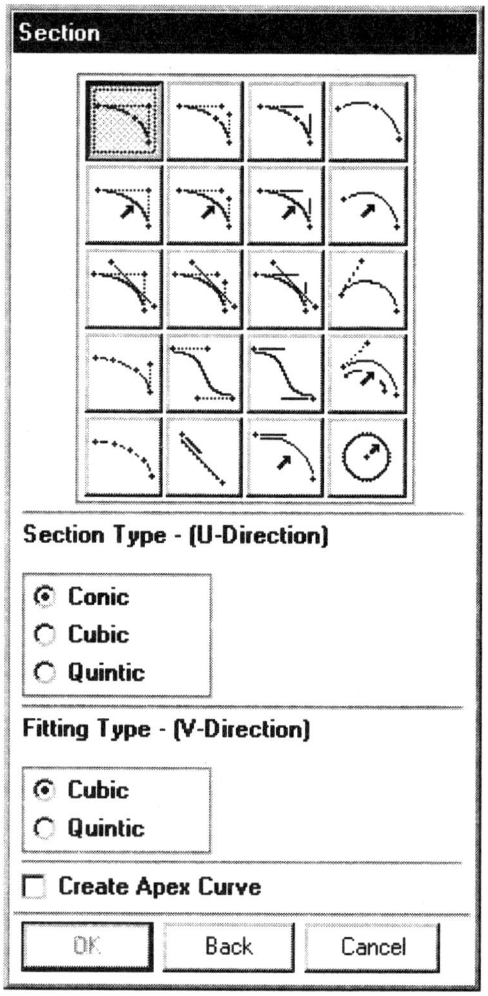

Section Definitions and Common Options:

Section creates a sheet body that has a cross section shape ofa conic. **Sections** are similar to conic curve creation and uses defining curves instead of defining points. Section sheet body also allows for slope control. The essential advantage of conic sheets is no reversal of curvature in the V direction of the sheet body.

Common Options are options that are common to many of the Section creation menus.

Start Edge String defines the start edge of the sheet body. The string must be tangent if multiple curves and edges are selected.

End Edge String defines the end edge of the sheet body. The string must be tangent if multiple curves and edges are selected.

Start Slope String defines a vector to control the slope along the start edge of the sheet body. The string must be tangent if multiple curves and edges are selected.

End Slope String defines a vector to control the slope along the end edge of the sheet body. The string must be tangent if multiple curves and edges are selected.

Shoulder Curve defines an interior curve that the sheet body is created through. The string must be tangent if multiple curves and edges are selected.

Apex Curve Definition:

The apex curve is created through the defining apex points at all of the implied conic sections of the sheet. It is the intersection of the tangent sheets at either side of the conic section.

The **Apex Curve** is commonly referred to as the common slope control curve or anchor curve. In situations where the apex curve must be computed, the option to create the apex curve is available. Activate the **Create Apex Curve** option in the **Section** dialog; it creates an apex curve along with the conic sheets where the **Apex Curve** is not defined.

Hilite:

When two curves define a line between them, the Sheet Body is tangent.

Rho Definition:

Some of the **Section** options allow the definition of a **Rho** value. The **Rho** value defines the elongation of the conic curve as well as what type the defining conic is created. **Rho** is the ratio of Y1 (the distance from the midpoint of the chord to the intersection of the conic curve with the chordal bisector) to Y2 (the distance from the chord midpoint to the apex). The value of this ratio lies between .001 and .999. A value between .001 and .499 generates an **ellipse**. The closer to 0, the flatter the **ellipse**. A value of .500 generates a **parabola** and a value between .501 and .999 generates a **hyperbola**. The closer to 1, the more peaked the resulting **hyperbola** is.

The Rho value can be specified using the following methods: **Constant**, **Least Tension**, and **General**.

Constant assigns a constant **Rho** value along the entire body.

Least Tension uses a least tension condition to determine the **Rho** from the geometry. A **Least Tension** conic has the minimum permissible degree of curvature, which satisfies the initial conditions and usually results in an **ellipse**. If the tangent value is the same on each side of the **conic**, this feature created is a **circle**.

General uses the **Law Subfunction** dialog to define the **Rho**. The **Law Subfunction** options are explained in the Advanced Curves Section.

Spine Definition:

A Spine Curve controls the shape of a conic sheet body. At each point on the **Spine Curve**, Unigraphics constructs a section plane normal to the tangent at that point. The section plane is then intersected with each of the control curves resulting in a series of points and slope control vectors that are used to construct a conic curve. The spine may be an analytic curve such as a line or an arc, or a complex b-curve. The complexity of the **Spine Curve** determines the complexity of the resultant conic sheet. If lines and arcs are used to create spine curves, the resultant conic sheets are simpler.
Selecting the control curves at either end does not affect the alignment direction of the curves. However, the **Spine Curve** selection is direction sensitive. The end you select becomes the start direction of the spine curve and determines the directional sense of the sheet. The spine can also determine the length of the resultant sheet. If the normal created at the start/end point of the spine is between the start/end points of the control curves, the spine normals limit the length of the sheet. If the normals are outside the end points of the control curves, the control curves determine the length of the sheet.

Section Type – (U Direction) controls the degree of translation in the U direction of the body. The degree of a cross section is taken normal to the defining strings of the section body. Conic creates an analytical defined shape, i.e., ellipse, hyperbola, parabola, or circle, in the U direction. Cubic creates a body with a degree of 3 in the U direction. Quintic creates a body with a degree of 5 in the U direction.

Filtering Type – (V Direction) controls the degree of translation in the V direction of the body. The degree of the body is taken along the defining strings.
Cubic creates a body with a degree of 3 in the V direction.
Quintic creates a body with a degree of 5 in the V direction.

Note: An error is generated if the spine curve has an "S" characteristic with a reversal of curvature. As Unigraphics creates the section plane along the spine, the intersection points generated on each control curve (start, shoulder, end curves) must be either increasing or decreasing along the curves. In other words, if the intersection points are found to "back track" on one of the control curves, then Unigraphics generates an error.

Ends – Apex – Shoulder:

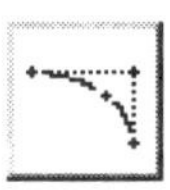 **Insert | Free Form Feature | Section**

This creates a conic sheet body by using three curves for the sheet to pass through and the fourth curve to serve as the apex. The apex curve determines the slope of the sheet at the first and last curve.

Exercise: Section creation with ends-apex-shoulder

1. Open **ff_10**.
2. Select **Insert** | **Free Form Feature** | **Section**.
3. Activate the type of **U-Direction** and **V-Direction**.
4.  Click **ends-apex-shoulder**.
5. Pick the curves for the **Start Edge String**.
6. Pick the curves for the **Shoulder String**.
7. Pick the curves for the **End Edge String**.
8. Pick the curves for the **Apex String**.
9. Pick the curves for the **Spine String**.
10. Click **OK** to create the **Sheet Body**.

Ends-Slope-Shoulder:

 **Insert** | **Free Form Feature** | **Section**

Create a conic sheet body through three curves and two slope control curves that are located at the start and end curve location. The interior curve that the **Sheet Body** is created through is the **Shoulder Curve**. The **Start Slope Control Curve** and **End Slope Control Curve** define a vector that controls the slope of the conic sheet at the **Start Edge Curve** and **End Edge Curve**. The vector is defined starting at the **Edge Curves** and pointing through the **Slope Control Curves**.

Exercise: Section creation with ends-slopes-shoulder

1. Open **ff_11**.
2. Select **Insert** | **Free Form Feature** | **Section**.
3. Activate the type of **U-Direction** and **V-Direction**.
4.  Click **ends-slopes-shoulder**.
5. Pick the curves for the **Start Edge String**.
6. Pick the curves for the **Start Slope Control String**.
7. Pick the curves for the **Shoulder String**.
8. Pick the curves for the **End Edge String**.
9. Pick the curves for the **End Slope Control String**.
10. Pick the curves for the **Spine String**.
11. Click **OK** to create the sheet body.

Fillet-Shoulder:

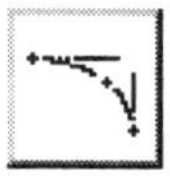 **Insert | Free Form Feature | Section**

This creates a conic sheet body that is tangent to two sets of faces and passes through a **Shoulder Curve**.

Exercise: Section creation with fillet-shoulder

1. Open **ff_12**.
2. Select **Insert | Free Form Feature | Section**.
3. Activate the type of **U-Direction** and **V-Direction**.
4.  Click **fillet-shoulder**.
5. Pick the face(s) for the **First Set of Faces**.
6. Pick the curves for the **String on First Set of Faces**.
7. Pick the curves for the **Shoulder String**.
8. Pick the face(s) for the **Second Set of Faces**.
9. Pick the curves for the **String on Second Set of Faces**.
10. Pick the curves for the **Spine String**.
11. Click **OK** to create the sheet body.

Three-Point-Arc:

 **Insert | Free Form Feature | Section**

This creates a conic sheet by using a Start Edge Curve, an Interior Curve, and an End Edge Curve. The cross-section of the sheet body is an arc.

Exercise: Section creation with three-point-arc

1. Open **ff_13**.
2. Select **Insert | Free Form Feature | Section**.
3. Activate the type of **U-Direction** and **V-Direction**.
4. 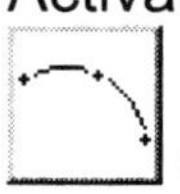 Click **three-point-arc**.
5. Pick the curves for the **Start Edge String**.
6. Pick the curves for the **First Interior String**.
7. Pick the curves for the **End Edge String**.

8. Pick the curves for the **Spine String**.
9. Click **OK** to create the sheet body.

Ends-Apex-Rho:

Insert | Free Form Feature | Section

This creates a conic sheet body using a start edge string, end edge string, an apex curve. The shape of the curve is controlled by your input for the Rho value.

Rho Definition Method:

Constant will allow you to enter a Rho value that is constant along the entire length of the conic sheet body.
Least Tension calculates the Rho value based on the curves selected as the **Start**, **End**, and **Apex** curves. The conic sheet body has minimal variations in curvature and is created close to a circular shape.
General defines the Rho value using the **Law Control** dialog.

Exercise: Section creation with ends-apex-Rho

1. Open **ff_14**.
2. Select **Insert | Free Form Feature | Section**.
3. Activate the type of **U-Direction** and **V-Direction**.
4. Click **ends-apex-Rho**.
5. Pick the curves for the **Start Edge String**.
6. Pick the curves for the **End Edge String**.
7. Pick the curves for the **Apex String**.
8. Pick the curves for the **Spine String**.
9. Click on the type of **Rho Definition Method**.
10. Enter a value for **Rho**.
11. Click **OK** to create the sheet body.

Ends-Slope-Rho:

Insert | Free Form Feature | Section

This creates a conic sheet body using a start edge string and end edge string. The slope is defined by a slope control curve at the start and end edge string. The shape of the curve is controlled by your input for the Rho value.

Rho Definition Method:

Constant will allow you to enter a Rho value that is constant along the entire length of the conic sheet body.
Least Tension calculates the Rho value based on the curves selected as the **Start**, **End**, and **Apex** curves. The conic sheet body has minimal variations in curvature and is created close to a circular shape.
General defines the Rho value using the **Law Control** dialog.

Exercise: Section creation with a start edge and end edge string

1. Open **ff_15**.
2. Select **Insert** | **Free Form Feature** | **Section**.
3. Activate the type of **U-Direction** and **V-Direction**.
4.  Click **ends-slopes-Rho**.
5. Pick the curves for the **Start Edge String**.
6. Pick the curves for the **Start Slope Control String**.
7. Pick the curves for the **End Edge String**.
8. Pick the curves for the **End Slope Control String**.
9. Pick the curves for the **Spine String**.
10. Click on the type of **Rho Definition Method**.
11. Enter a value for **Rho**.
12. Click **OK** to create the sheet body.

Fillet-Rho:

 **Insert** | **Free Form Feature** | **Section**

This creates a conic sheet body that is tangent to two sets of faces. The conic created forms a blend tangent to the first set of faces, starting at the string on the first set of faces and is tangent to the second set of faces, ending at the string on the second set of faces. The shape of the curve is controlled by your input for the Rho value.

Rho Definition Method:

Constant will allow you to enter a Rho value that is constant along the entire length of the conic sheet body.
Least Tension calculates the Rho value based on the curves selected as the **Start**, **End**, and **Apex** curves. The conic sheet body has minimal variations in curvature and is created close to a circular shape.
General defines the Rho value using the **Law Control** dialog.

Exercise: Section creation with tangent to two faces

1. Open **ff_16**.
2. Select **Insert | Free Form Feature | Section**.
3. Activate the type of **U-Direction** and **V-Direction**.
4.  Click **fillet-Rho**.
5. Pick the face(s) for the **First Set of Faces**.
6. Pick the curves for the **String on First Set of Faces**.
7. Pick the face(s) for the **Second Set of Faces**.
8. Pick the curves for the **String on Second Set of Faces**.
9. Pick the curves for the **Spine String**.
10. Click the type of **Rho Definition Method**.
11. Enter a value for **Rho**.
12. Click **OK** to create the sheet body.

Two Point Radius:

 **Insert | Free Form Feature | Section**

This creates a circular sheet body between two curves. The sheet body is a constant radius that is created in a counter-clockwise direction. The radius must be greater than half the distance between the two curves.

Use Radius Law:

When the **Section Radius Data** dialog displays this option is available to define the radius. Selecting the **Use Radius Law** button displays the Law **Control dialog to define the radius.**

Exercise: Section creation with two-points-radius

1. Open **ff_17**.
2. Select **Insert | Free Form Feature | Section**.
3. Activate the type of **U-Direction** and **V-Direction**.
4.  Click **two-points-radius**.
5. Pick the curves for the **Start Edge String**.
6. Pick the curves for the **End Edge String**.
7. Pick the curves for the **Spine String**.
8. Enter a value in the text box for the **Radius**.
9. Click **OK** to create the sheet body.

Ends-Apex-Hilite:

 **Insert | Free Form Feature | Section**

This creates a conic sheet body by using a startand end edge string for the sheet to created from and set of curves to serve as the apex. The apex curve determines the slope of the sheet at the first and last curve. A start and end set of hilite curves are used to create a line in between them for the conic sheet body to be tangent to.

Exercise: Section creation with ends-apex-hilite

1. Open **ff_18**.
2. Select **Insert | Free Form Feature | Section**.
3. Activate the type of **U-Direction** and **V-Direction**.
4.  Click **ends-apex-hilite**.
5. Pick the curves for the **Start Edge String**.
6. Pick the curves for the **End Edge String**.
7. Pick the curves for the **Apex String**.
8. Pick the curves for the **Hilite Start String**.
9. Pick the curves for the **Hilite End String**.
10. Pick the curves for the **Spine String**.
11. Click **OK** to create the sheet body.

Ends-Slope-Hilite:

 **Insert | Free Form Feature | Section**

Create a conic sheet body by using a start and end edge string for the sheet to create from and a start and end slope string to define the slope. A start and end set of hilite curves are used to create a line in between them for the conic sheet body to be tangent to.

Exercise: Section creation with ends-slopes-hilite

1. Open **ff_19**.
2. Select **Insert | Free Form Feature | Section**.
3. Activate the type of **U-Direction** and **V-Direction**.

4.  Click **cnds-slopes-hilite**.
5. Pick the curves for the **Start Edge String**.
6. Pick the curves for the **Start Slope Control String**.
7. Pick the curves for the **End Edge String**.
8. Pick the curves for the **End Slope Control String**.
9. Pick the curves for the **Hilite Start String**.
10. Pick the curves for the **Hilite End String**.
11. Pick the curves for the **Spine String**.
12. Click **OK** to create the sheet body.

Fillet-Hilite:

 **Insert | Free Form Feature | Section**

This creates a conic sheet body that is tangent to two sets of faces. The conic created forms a blend tangent to the first set of faces, starting at the string on the first set of faces and is tangent to the second set of faces, ending at the string on the second set of faces. A start and end set of hilite curves are used to create a line in between them for the conic sheet body to be tangent to.

Exercise: Section creation with fillet-hilite

1. Open **ff_20**.
2. Select **Insert | Free Form Feature | Section**.
3. Activate the type of **U-Direction** and **V-Direction**.
4.  Click **fillet-hilite**.
5. Pick the face(s) for the **First Set of Faces**.
6. Pick the curves for the **String on First Set of Faces**.
7. Pick the face(s) for the **Second Set of Faces**.
8. Pick the curves for the **String on Second Set of Faces**.
9. Pick the curves for the **Hilite Start String**.
10. Pick the curves for the **Hilite End String**.
11. Pick the curves for the **Spine String**.
12. Click **OK** to create the sheet body.

Ends-Slope-Arc:

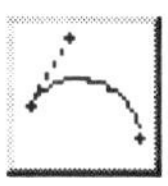 **Insert | Free Form Feature | Section**

Create a circular conic sheet body using a **Start** and **End Edge String**. Only slope on the **Start Edge String** is controlled. The slope is defined from the **Start Edge**

String through the **Start Slope Control Curve**. The cross-section of the sheet body is an arc.

Exercise: Section creation with ends-slope-arc

1. Open **ff_21**.
2. Select **Insert | Free Form Feature | Section**.
3. Activate the type of **U-Direction** and **V-Direction**.
4.  Click **ends-slope-arc**.
5. Pick the curves for the **Start Edge String**.
6. Pick the curves for the **Start Slope Control String**.
7. Pick the curves for the **End Edge String**.
8. Pick the curves for the **Spine String**.
9. Click **OK** to create the sheet body.

Four-Points-Slope:

 **Insert | Free Form Feature | Section**

This creates a conic sheet body using four curves and defining a slope along the **Start Edge String**. A vector starting at the Start Edge String and pointing through the Start Slope Control String determines the slope of conic sheet body.

Exercise: Section creation with four-points-slope

1. Open **ff_22**.
2. Select **Insert | Free Form Feature | Section**.
3. Activate the type of **U-Direction** and **V-Direction**.
4.  Click **four-points-slope**.
5. Pick the curves for the **Start Edge String**.
6. Pick the curves for the **Start Slope Control String**.
7. Pick the curves for the **First Interior String**.
8. Pick the curves for the **Second Interior String**.
9. Pick the curves for the **End Edge String**.
10. Pick the curves for the **Spine String**.
11. Click **OK** to create the **Sheet Body**.

Ends-Slopes-Cubic:

 **Insert | Free Form Feature | Section**

Create an S-shaped (cubic) sheet body between two curves. A vector starting at the Start and End Edge String and pointing through the Start and End Slope Control String determines the slope of the conic sheet body.

Exercise: Section creation using ends-slopes-cubic

1. Open **ff_23**.
2. Select **Insert | Free Form Feature | Section**.
3. Activate the type of **U-Direction** and **V-Direction**.
4.  Click **ends-slopes-cubic**.
5. Pick the curves for the **Start Edge String**.
6. Pick the curves for the **Start Slope Control String**.
7. Pick the curves for the **End Edge String**.
8. Pick the curves for the **End Slope Control String**.
9. Pick the curves for the **Spine String**.
10. Click **OK** to create the sheet body.

Fillet Bridge:

 **Insert | Free Form Feature | Section**

This creates a surface between two sets of faces. Control the shape of a bridge sheet body with "free-hand" surface styling. Control the shape of a bridge sheet body using the parent surfaces' isoperimetric cross sections.

Bridge Section Options:

Match Tangents creates a sheet body with its sides matching the tangency of the faces selected to build the sheet body between. A spine curve is optional for this option.
Match Curvatures creates a sheet body with its sides matching the curvature of the faces selected to build the sheet body between. A spine curve is optional for this option.

Inherit Shape allows the selection of a string of curves to control the shape at the start and end of the sheet body. A spine curve must be selected for this option.

Reverse direction flips the direction of the bridge surface's transition off the parent surfaces. When you click Reverse Direction, the edge from the first set of faces flips. The second time you click Reverse Direction the other edge flips. This continues to alternate as you select Reverse Direction.

1st wall match and 2nd wall match control the bridge surface's transition from the first and second set of faces. **Tangency** creates a flatter continuity and **Curvature** creates continuity with more shape.

Control region helps industrial designers when aesthetically pleasing results require changes in shape to the surface from end to end. This setting controls how much of the surface changes when using the Bridge Depth and Bridge Skew dynamic scroll bars. Bridge Depth controls the Rho of the conic surface and Bridge Skew controls the location of maximum curvature.

Entire manipulates the entire bridge sheet when using the dynamic scroll bars.

Start only manipulates the start of the bridge and tapers out to zero change as it reaches the end. The end you selected first when selecting strings determines the start position.

End only manipulates the end of the bridge and tapers out to zero change as it reaches the start.

Stiffness control controls the continuity, degree, and complexity of the bridge surface. Stiffness Control is only available if you set 1st Wall Match and 2nd Wall Match to Curvature.

Auto attempts to create the smoothest possible 5-degree surface.

Low creates a 5-degree surface with more shape but little control over continuity.

High creates a 7-degree surface with smooth continuity and more shape. High can sometimes result in too much shape if the parent surfaces have a lot of curvature.

Flow Direction:

Flow Direction on 1st Wall and **Flow Direction on 2nd Wall** control the transition of the isoperimetric cross-sections of the bridge surface to the parent surfaces. These options are only available if a spine string is not selected.

Not Specified places the cross-section of the bridge sheet along a plane approximating as closely as possible the normal of each parent surface.

Perpendicular places the cross section of the bridge sheet so that either side or both sides are exactly perpendicular to the slope of the matching curve on that wall.

Iso Line U and **Iso Line V** align the cross section with the isoperimetric curve on each wall respectively.

Exercise: Creating a fillet bridge with match tangents and match curvature

1. Open **ff_24**.
2. Select **Insert | Free Form Feature | Section**.
3.  Click **fillet-bridge**.
4. Click **Match Tangents** or **Match Curvatures**.
5. Pick the first set of faces from the Graphics window.
6. Click **OK**.
7. Pick the first set of face strings in the Graphics window.
8. Click **OK**.
9. Pick the second set of faces in the Graphics window.
10. Click **OK**.
11. Pick the second set of face strings in the Graphics window.
12. Click **OK**.
13. Either pick the geometry in the Graphics window to define the spine string and/or click **OK** to display the **Shape Control** dialog.
14. Change settings as needed to create the surface.
15. Click **OK**.

Exercise: Creating a fillet bridge with inherit shape

1. Open **ff_25**.
2. Select **Insert | Free Form Feature | Section**.
3.  Click **fillet-bridge**.
4. Click **Inherit**.
5. Pick the first set of faces in the Graphics window.
6. Click **OK**.
7. Pick the first set of face strings in the Graphics window.
8. Click **OK**.
9. Pick the second set of faces in the Graphics window.
10. Click **OK**.
11. Pick the second set of face strings in the Graphics window.
12. Click **OK**.
13. Pick the curves for the **Starting Shape Curve String** in the Graphics window.
14. Click **OK**.

15. Pick the curves for the **Ending Shape Curve String** in the Graphics window.
16. Click **OK**.
17. Pick the spine curve string in the Graphics window.
18. Click **OK** to create the sheet body.

Note: Be sure to pick the First and Second Set String in sequential order to avoid getting a twisted surface.

Point-Radius-Angle-Arc:

 **Insert | Free Form Feature | Section**

This creates a circular sheet body that is tangent to a face and passes through an edge. Your inputs for the radius and angle control the size and end angle of the circular sheet body.

Section options control the size of the Radius, Angle and which direction the sheet body is created from the selected face.

A **Radius** value can be entered in the text box for a constant radius along the string. Clicking the **Use Radius Law** button displays the **Law Control** dialog.

An **Angle** value can be entered in the text box for a constant angle for the end of the sheet body from the string. Clicking the **Use Angle Law** button displays the **Law Control** dialog.

Reverse Side of Face flips the direction of the vector that is displayed.

Exercise: Section creation with point-radius-angle-arc

1. Open **ff_26**.
2. Select **Insert | Free Form Feature | Section**.
3. Activate the type of **U-Direction** and **V-Direction**.
4.  Click **point-radius-angle-arc**.
5. Pick the face(s) for the **First Set of Faces**.
6. Pick the curves for the **String on First Set of Faces**.
7. Pick the curves for the **Spine String**.
8. Enter a value for the **Radius**.

9. Enter a value for the **Angle**.
10. Click **Reverse Side of Face** to flip the vector in the Graphics window, if needed.
11. Click **OK** to create the sheet body.

Note: The sheet body can have a span of less than 180 degrees. However the span cannot pass through zero. In other words, the angle can be from 0 to 180 or-180 to 0 but not -15 to 180. The default position of the sheet is in the direction of the face normal, or it can be flipped to the opposite side of the tangent face.

Five Points:

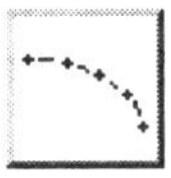 **Insert | Free Form Feature | Section**

This creates a conic sheet by using five curves as the control curves of the sheet body. This is similar to creating a general conic through five points. Unigraphics best fits a conic sheet body through the curves with the resultant shape being an ellipse, parabola or a hyperbola.

Exercise: Section creation with five-points

1. Open **ff_27**.
2. Select **Insert | Free Form Feature | Section**.
3. Activate the type of **U-Direction** and **V-Direction**.
4. 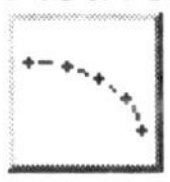 Click **five-points**.
5. Pick the curves for the **Start Edge String**.
6. Pick the curves for the **First Interior String**.
7. Pick the curves for the **Second Interior String**.
8. Pick the curves for the **Third Interior String**.
9. Pick the curves for the **End Edge String**.
10. Pick the curves for the **Spine String**.
11. Click **OK** to create the **Sheet Body**.

Linear-Tangent:

 **Insert | Free Form Feature | Section**

This creates a sheet body between an existing set of faces and a string, which is defined by a set of curves or edges. The body created is tangent from the stat string to the set of faces. Unigraphics determines the location of the tangency occurrence by the nearest location that tangency can occur by projecting the curve or edge along a vector.

Exercise: Section creation with linear-tangent

1. Open **ff_28**.
2. Select **Insert | Free Form Feature | Section.**
3. Activate the type of **U-Direction** and **V-Direction.**
4.  Click **linear-tangent**.
5. Pick the face(s) for the **Set of Tangent Faces**.
6. Pick the curves for the **Start String**.
7. Pick the curves for the **Spine String.**
8. Click **OK** to create the sheet body.

Note: A supporting face may have to be picked if the tangent faces selected are too small for all the points of the start string to project onto. Creating an extension sheet is an option that solves this problem.

Circular-Tangent:

 **Insert | Free Form Feature | Section**

This creates a circular sheet body between an existing set of faces and a string, which is defined by a set of curves or edges. The body created is tangent from the start string to the set of faces. You have the option to create the sheet body as a fillet or cover arc.

Section options:

There are two Circular Creation Types: **Fillet Arc** and **Cover Arc** (concave or convex). A radius value is necessary for both of these types. Type in a constant value or click the **Use Radius Law** button to display the **Law Control** dialog. The following figure illustrates the two creation types.

Exercise: Section creation circular-tangent

1. Open **ff_29**.
2. Select **Insert | Free Form Feature | Section**.
3. Activate the type of **U-Direction** and **V-Direction**.
4. 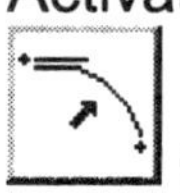 Click **circular-tangent**.
5. Pick the face(s) for the **Set of Tangent Faces**.
6. Pick the curves for the **Start String**.
7. Pick the curves for the **Spine String**.
8. Activate the **Circle Creation Type** needed and enter a value for the **Radius**.
9. Click **OK** to create the sheet body.

Note: If the Start String is longer than the face being projected onto an error message displays saying "Failed to construct section at start of the limit curve."

Circle:

Insert | Free Form Feature | Section

This option creates a circular section body along a guide string. The size of the radius is controlled by your input and can be either constant or law controlled.

Section radius data maintains a constant radius along the guide string. Clicking **Use Radius Law** displays the Law Control dialog. If **Radius Law** is used, the radius value cannot be 0 except at the ends of the section body.

Orientation Curve String:

When using the optional **Orientation Curve String,** the section body ends where the last line, normal to the spine, can be created and also intersects the Orientation Curve String. The cross section of the section body is still controlled by the **Spine** curve.

Exercise: Section creation with circle

1. Open **ff_30**.
2. Select **Insert | Free Form Feature | Section**.
3. Activate the type of **U-Direction** and **V-Direction**.
4. Click **circle**.
5. Pick the curves for the **Guide String**.
6. Pick the curves for the **Orientation Curve String**, (optional).
7. Pick the curves for the **Spine String**.
8. Enter a value for the **Radius**.
9. Click **OK** to create the sheet body.

Note: The body is created along the guide string with the cross-section normal to the spine string at all defining points. If the Spine String is shorter than the guide string, the body ends where a plane, normal to the spine and at the end of the spine, intersects the guide string.

Extension Sheet Body:

 **Insert | Free Form Feature | Extension**

The length of the extension can be either a specified distance or a percentage of the existing sheet body. The extension sheet is a separate sheet from the base sheet and is associative to the length and tolerance of the base sheet.

Tangential creates an extension sheet that is tangent to an edge or corner of a base sheet. The edges picked must be original edges and not the results of a trimming operation.

> **Fixed Length** creates a tangential extension of a defined length.
>
> **Percentage** creates a tangential extension as a percentage of the total U or V direction of the base sheet.
>
> **Edge Extension** creates a sheet body originating from the edge selected. The resultant sheet is tangent and has the same width as the edge selected on the base sheet.
>
> **Corner Extension** creates a sheet body originating from the corner selected. After you select **Corner Extension**, Unigraphics prompts you to pick the face and the corner you wish to extend.
>
> A U and V vector appears on the corner of the base sheet selected. The percentage for both the U and V and directions are taken from the base sheet.

Normal to surface creates an extension sheet normal to a face or sheet body through a curve that lies on the face or sheet body. The length of the curve and the length value determines the size of the extension sheet. The face selected determines the normal direction of the extension sheet body.

> A positive **Length** value creates a sheet in the direction of the vector. A negative value creates the sheet in the opposite direction.

Angled creates an extension sheet at an angle to a face or sheet body through a curve that lies on the face or sheet body. The length of the curve, the angle value, and the length value determines the size of the extension sheet
Unigraphics displays two vectors to determine the positive direction of the angle. Positive direction is determined from the first vector (0), rotating towards the second (90), tangent to the sheet, and normal to the sheet.

Circular creates an extension sheet that is tangent to an edge and matches the curvature of the base sheet at that edge.

Fixed Length creates a circular extension of a defined length.

Percentage creates a circular extension as a percentage of the total U or V direction of the base sheet.

Law controlled creates an unparameterized extension sheet with the length and angle controlled by a Law option.
By selecting **Measured Angle From Plane**, you can create the extension at an angle from the specified plane. If you do not select a plane, Unigraphics creates the extension sheet at an angle to the face selected.

Exercise: Creating a fixed length tangential or circular extension sheet body

1. Open **ff_31**.
2. Select **Insert | Free Form Feature | Extension**.
3. Click **Tangential**, or **Circular**.
4. Click **Fixed Length**.
5. Pick the sheet body in the Graphics window.
6. Pick the edge from which the extension sheet extends.
7. Enter a value for the **Length**.
8. Click **OK** to create the sheet body.

Exercise: Creating a tangential sheet body with percentage

1. Open **ff_32**.
2. Select **Insert | Free Form Feature | Extension**.
3. Click **Tangential**.
4. Click **Percentage**.
5. Click **Edge Extension**.
6. Pick a sheet body in the Graphics window.
7. Pick the edge from which you want the extension sheet to extend.
8. Enter a value for the **Percentage**.
9. Click **OK** to create the sheet body.

Exercise: Creating a corner extension sheet body

1. Open **ff_33**.
2. Select **Insert | Free Form Feature | Extension**.
3. Click **Tangential**.
4. Click **Percentage**.
5. Click **Corner Extension**.
6. Pick a sheet body in the Graphics window.

7. Pick near the corner of the sheet body from which you want the extension sheet to extend.
8. Enter a value for **U** and **V**.
9. Click **OK** to create the sheet body.

Exercise: Creating a circular extension sheet body with percentage

1. Open **ff_34**.
2. Select **Insert | Free Form Feature | Extension**.
3. Click **Circular**.
4. Click **Percentage**.
5. Pick a sheet body in the Graphics window.
6. Pick the edge from which you want the extension sheet to extend.
7. Enter a **Percentage** value.
8. Click **OK** to create the sheet body.

Exercise: Creating a normal to surface extension surface

1. Open **ff_35**.
2. Select **Insert | Free Form Feature | Extension**.
3. Click **Normal To Surface**.
4. Pick the sheet body in the Graphics window.
5. Pick the curve that lies on the sheet body from which you want the extension sheet to extend.
6. Type in the value for the **Length**.
7. Click **OK** to create the sheet body.

Exercise: Creating an angled extension surface

1. Open **ff_36**.
2. Select **Insert | Free Form Feature | Extension**.
3. Click **Angled**.
4. Pick the sheet body in the Graphics window.
5. Pick the curve that lies on the sheet body from which you want the extension sheet to extend.
6. Type in the values for the **Length** and **Angle**.
7. Click **OK** to create the sheet body.

Exercise: Creating a law controlled surface

1. Open **ff_37**.
2. Select **Insert | Free Form Feature | Extension**.
3. Click **Law Controlled**.
4. Pick a sheet body in the Graphics window.
5. Click **OK**.
6. Pick a curve string in the Graphics window.

7. Click **OK**.
8. Pick a spine string or click **OK** if you do not desire one.
9. Specify a law option for length.
10. Specify a law option for angle.
11. Click **OK** to create the sheet body.

Notes: Use the same percentage on adjacent edge and corner extension sheets to match the new sheets exactly.

Angled Extension sheets created at a 0-degree angle create a tangential extension sheet. You can use this on trimmed base sheets where tangential extensions fail.

The base curve used to define tangency of the circular extension must be a non-trimmed edge of the base sheet. The length specified must be small enough so as not to create a full circular extension on any part of the extension sheet or a self-intersecting sheet may result.

Surface Fillets:

 **Insert | Free Form Feature | Fillet**

This creates an unparameterized sheet tangent to two sheets or faces that is either circular or conical in a cross-sectional shape.

Creating a fillet or curve determines whether you create a fillet sheet and/or curves.

Create Fillet switches between **Yes** and **No**. **Yes** indicates that a fillet sheet will be created. **No** indicates that a fillet sheet will not be created.

Create Curve switches between **Yes** and **No**. **Yes** indicates that a curve that represents the fillet sheet's centerline will be created. **No** indicates that a centerline will not be created.

Cross section type determines the shape of the fillet sheet body.

Circular creates a fillet sheet where the cross-section is an arc. The input value is a radius.

Conic creates a fillet sheet where the cross section behaves more lke a General Conic. The shape of the fillet depends upon the input values given for the conic.

Fillet type determines the transition of the fillet sheet body from the start position to the end position.

Constant creates a constant radius fillet sheet between the two selected faces for **Circular** and a constant conical cross section for **Conic**.

Linear creates a circular fillet sheet between the two selected faces that has a different radius at each end of the fillet sheet. The translation between the two different radii is a constant transition.
The **Linear** option allows you to create a conic fillet sheet between the two selected faces that has a different radius, ratio, and Rho value at each end of the fillet sheet.

S-Shaped works exactly as **Linear** with the exception of the way in which the fillet sheet blends between the start and end locations. The figure illustrates the difference between the two types. This transition is similar to the cubic transition of a swept feature.

General is only available when you select a spine curve. It creates a variable radius fillet sheet between the two selected faces that can have a different radius at any specified point along the spine curve.

Rho inputs become available when you select **Conic** as your fillet type.

The **Same As Fillet Type** option creates a **Conic Fillet** as specified by the fillet type chosen: **Constant, Linear, S-Shaped,** or **General**.
The Least Tension option uses the geometry selected to determine what type of conic is created. You usually create an elliptical filet with this method.

You input radius, ratio, and Rho values to define the conic fillet size.

Radius determines the size of the fillet.

Ratio determines how oblong the fillet is. The **Ratio** is the ratio of the distances from the faces to the centerline curve. A ratio less than one sets the centerline curve further from the first face. A ratio equal to one sets the centerline at equal distances to the faces, and a ratio greater than 1 sets the centerline curve further from the second face.

Rho determines what type of fillet you create. This relates back to the standard shapes created from a varying Rho value:
.001< RHO <.500 is an ellipse
RHO =.500 is a parabola
.500 < RHO < .999 is a hyperbola.

Fillet start point is available if you do not select a spine curve. These determine the start and end positions and shapes.

Limit Point displays the **Point Constructor** dialog. Pick a location for the fillet sheet to start or end.

Limit Face allows you to pick a face of a solid or another sheet body as a limiting feature.

Limit Plane displays the **Plane Subfunction** dialog for you to define a plane as a limiting feature.

Exercise: Creating a circular fillet with a spine

1. Open **ff_38**.
2. Select **Insert | Free Form Feature | Fillet**.
3. Pick a face in the Graphics window.
4. Click **Yes** to accept the direction or **No** to reverse the direction.
5. Pick the second face in the Graphics window.
6. Click **Yes** to accept the direction or **No** to reverse the direction.
7. Pick a curve in the Graphics window.
8. Make sure that **Create Fillet –Yes**.
9. Click **Circular**.
10. Choose a shape option.
11. Pick a point at which you want to start the fillet surface.
12. Enter a **Radius** value
13. Click **Yes** or **No** depending on the direction of the vector.
14. Pick a point at which you want to end the fillet surface.
15. Enter a **Radius** value.
16. Click **OK**.
17. Click **Yes** to create the fillet sheet.

Exercise: Creating a conical fillet with a spine

1. Open **ff_39**.
2. Select **Insert | Free Form Feature | Fillet**.
3. Pick a face in the Graphics window.
4. Click **Yes** to accept the direction or **No** to reverse the direction.
5. Pick the second face in the Graphics window.
6. Click **Yes** to accept the direction or **No** to reverse the direction.
7. Pick a curve in the Graphics.
8. Make sure that **Create Fillet -Yes** is active.
9. Click **Conic**.
10. Choose a shape option.
11. Choose a Rho option.
12. Pick a point from which you want to start the fillet surface.
13. Enter the desired values.
14. Click **Yes** or **No** depending on the direction of the vector.
15. Pick a point from which you want to end the fillet surface.

16. Enter the desired values and click **OK**.
17. Click **Yes** to create the fillet sheet.

Exercise: Creating a circular filler without a spine

1. Open **ff_40**.
2. Select **Insert | Free Form Feature | Fillet**.
3. Pick a face in the Graphics window.
4. Click **Yes** to accept the direction or **No** to reverse the direction.
5. Pick the second face in the Graphics window.
6. Click **Yes** to accept the direction or **No** to reverse the direction.
7. Click **OK**.
8. Make sure that **Create Fillet -Yes** is active.
9. Click **Circular**.
10. Choose a shape option.
11. Click a start position.
12. Pick a point, plane, or face from which you want to start the fillet surface.
13. Enter a **Radius** value and click **OK**.
14. Click **Yes** or **No** depending on the direction of the vector.
15. Click an end position option.
16. Pick a point, plane, or face from which you want to end the fillet surface.
17. Enter a **Radius** value and click **OK**.
18. Click **Yes** to create the fillet sheet.

Exercise: Creating a conical fillet without a spine

1. Open **ff_41**.
2. Select **Insert | Free Form Feature | Fillet**.
3. Pick a face from the Graphics window.
4. Click **Yes** to accept the direction or **No** to reverse the direction.
5. Pick the second face in the Graphics window.
6. Click **Yes** to accept the direction or **No** to reverse the direction.
7. Click **OK**.
8. Make sure that **Create Fillet -Yes** is active.
9. Click **Circular**.
10. Choose a shape option.
11. Choose a Rho option.
12. Click a start position option.
13. Pick a point, plane, or face from which you want to start the fillet surface.
14. Enter the desired values and click **OK**.
15. Click **Yes** or **No** depending on the direction of the vector.
16. Click an end position option.
17. Pick a point, plane, or face from which you want to end the fillet surface.
18. Enter the desired values and click **OK**.
19. Click **Yes** to create the fillet sheet.

Notes: In the **Fillet Options** dialog, both options cannot be set to **No**. If **Create Fillet** is set to **No**, Unigraphics automatically changes **Create Curve** to **Yes**.

You can use **Face Blend** to create fillet sheets. You can edit the parameters of a face blend.

Offset:

 **Insert | Free Form Feature | Offset**

This creates an offset sheet body from a face of a solid of from another sheet body.

Unigraphics creates the offset sheet by projecting points along the face normal. The points are offset a specified distance and define the offset sheet body. The offset sheet remains associative to the original face so that any edits to the original face will update the offset sheet.

Create a variable offset sheet. You have to specify unique offset distances at any four points on the base face/sheet to create the variable offset.

Distance specifies the total offset or variable offset of a new sheet. The distance is positive in the direction of the arrow but you can enter negative values to create the offset sheet in the other direction.

Exercise: Creating an offset surface

1. Open **ff_42**.
2. Select **Insert | Free Form Feature | Offset**.
3. Pick the face to offset.
4. Click **OK**.
5. Enter a **Distance value**.
6. If desired, change the value in the **Edge tolerance** text box.
7. Click **OK** to create the offset face.

Exercise: Creating a variable offset surface

1. Open **ff_43**.
2. Select **Insert | Free Form Feature | Offset**.
3. Pick the face to offset.
4. Click **OK**.
5. Click **Variable**.
6. Pick a corner point on the face in the Graphics window.
7. Enter a **Distance** value.
8. Click **OK**.

9. Repeat the last three steps until you picked all four corners.

Note: An unexpected offset body may result if the base face has poorly defined normals (i.e., normals that reverse or become nil) or if the offset distance is large enough to produce a self-intersecting offset body.

Offset Multiple Faces:

 **Insert | Free Form Feature | Offset**

Offset multiple faces in one step. The faces do not have to be from the same sheet or solid body. Individual offset features are created, allowing you to edit them individually at a later time.

Specify interior position is only available when the faces to offset are not adjacent. It allows you to define the normal vector of the selected faces by using the **Point Constructor** dialog to specify a point that defines the interior of each face. The side of each face pointing toward the specified point is the interior side. The side opposite to this is the exterior side, which defines the normal vector.

Use existing normals is available when the faces to offset are not adjacent. It allows you to use the existing normals of each face as the normal vector for the offset. The normal vector defines the direction of offset and the normal vector of the resulting face(s).

Redisplay normals is available when the faces to offset are not adjacent and after you define the normal vectors. In the **Reverse Normals** dialog, pick any faces needing the normal vector reversed. After selecting the faces, you must refresh the Graphics window. After refreshing the screen, use this button to redisplay the normal vectors.

Exercise: Offset multiple adjacent faces

1. Open **ff_44**.
2. Select **Insert | Free Form Feature | Offset**.
3. Pick the faces to offset.
4. Click **OK**.
5. Enter a **Distance** value.
6. If desired, change the value in the **Edge tolerance** text box.
7. Click **OK** to create the offset faces.

Exercise: Offsetting multiple non-adjacent faces

1. Open **ff_45**.

2. Select **Insert | Free Form Feature | Offset**.
3. Pick the faces to offset.
4. Click **OK**.
5. If desired, click **Specify Interior Position**. Pick a point to define the part side of the face. The normal direction points away from the interior. The **Reverse Normals** dialog displays and the normal vectors highlight.
   Or
   if desired, click **Use Existing Normals**.
6. If desired, pick any of the selected faces and click **OK** to reverse the normal vector(s).
7. To view the new normal vectors, right-click and select **Refresh** in the Graphics pop-up menu then click **Redisplay Normals**.
8. Click **OK**.
9. Enter a **Distance** value.
10. If desired, change the value in the **Edge tolerance** text box.
11. Click **OK** to create the offset faces.

Quilt:

You can combine a group of sheet bodies into one sheet body with one face to simplify operations requiring these faces. **Quilt** works well as long as the face edges meet within a limited tolerance.

The **Quilt** free form feature approximates any number of smaller sheets into one single sheet. Unigraphics creates a single B-surface that approximates a four-sided region lying on several existing sheets/faces. You create this B-surface by projecting points from a driver surface, either along a vector or along the surface normals of the driver, onto multiple sheets/faces. This region then defines the new quilt sheet body.

The new sheet is associated to the sheets/faces used to define it. The sheet updates to the modifications as long as the parametric link exists.

The Quilt feature is useful for applications that require a single face. You can use this feature to replace multiple faces with a single face. This condition arises when you sew two tangent sheets together and then use the resultant sheet to trim a body. After the trimming operation, the tangency seam scar leaves a resultant edge. If you create a quilt to replace the sewn sheet, there not be a seam edge left after the trim.

Any time you create a cosmetic part, you want to keep the surface as "clean" as possible. You do not want any blemishes caused by an edge to appear on your detail part. Manufacturing processes work faster if they only have to machine one face also. Quilt eliminates these problems; in addition, the resultant quilt surface is a 3-degree B-surface.

There are some limitations to the quilt algorithm. Because it projects points from the driver to the target sheets/faces, any target that becomes close to parallel to the projection vector causes problems.

The approximation handles target sheets/faces that do not match or have gaps, but if the targets have large overlaps or sharp ridges this could cause problems as well.

Insert | Free Form Feature | Quilt

This creates a single B-surface that approximates a four-sided region lying on several existing sheets/faces.

Driver type is the sheet or curves used to generate the points projected onto the original collection of sheets to define the quilt surface. In the **Driver Type** area of the **Quilt** dialog, there are two options to define a driver, **Mesh of Curves,** and **B-surface.** Use **Self-Refit** to decrease a high degree surface to a third degree B-surface.

> **Mesh of Curves** creates a driver internally to define the points needed to project. This is still a B-surface sheet but it is defined as though it was a Through Curve Mesh sheet. The same processes and requirements used in the creation of a Through Curve Mesh sheet are applicable. Unigraphics needs at least two (but less than 50) primary and secondary curves to construct the driver. The curves used to construct the mesh must be single curves and not a string of curves. The outer most primary and secondary curves are used to define the boundaries of the quilt. They must intersect each other within tolerance and fall entirely within the boundaries of the sheets to quilt. This B-surface is invisible and only exists until Unigraphics creates the quilt surface.
> This option is especially useful if you have an existing planar or extruded surface that you want to use as a driver. Simply select the surface edges instead of the surface itself.
>
> **B-surface** uses an existing surface as a driver surface. This surface must be a B-surface; a planar surface does not work.

Self-Refit is a third option available to define a surface driver called **Self-Refit**, however, this is not a driver selection. This option allows you to approximate a single B-surface with a quilt surface. This is useful for converting high degree surfaces to a Bi-Cubic polynomial surface.

Selecting the Along Fixed Vector option and using the Vector Constructordialog can define the **Projection Type**, or selecting Along Driver Normals can specify it.

Along Fixed Vector projects all the points in a parallel direction specified by the defined vector.

Along Driver Normals projects every point of the driver sheet along its normal on the sheet. You can define the **Projection Limit** to limit the distance a point projects. This helps Unigraphics determine which surface to use if it finds more than one intersection. This condition can arise if the sheet is cylindrical or if a number of sheets are used and there are overlap conditions.

Tolerances:

Tolerances are an important part of creating a quilt surface. If the tolerances are set too tight, a large number of patches on the quilt surface can result and you may be worse off than you were to begin with.

There are four tolerance values to input: **Inside Distance**, **Inside Angle**, **Edge Distance** and **Edge Angle**. The distance tolerances default to the modeling distance tolerance, .0254. The recommended range for the distance values is .0254 to .100.

Inside Distance permits you to specify an allowable distance deviation between corresponding points of the target surfaces and the quilt surface within the interior of the quilt surface. The default value is set through the **Modeling Preferences** dialog under Distance Tolerance.

Edge Distance permits you to specify an allowable distance deviation between the edges of the quilt surface and corresponding points on the target surface faces. We recommend keeping this tolerance tight if you need to sew the quilt surface to other surfaces.

Inside Angle allows you to specify an angle deviation between corresponding points on the interiors of both the target surfaces and the quilt surface. The default value is 90-degrees, which allows for any deviation.

The default value is the maximum of 90-degrees, which turns off the check.

Edge Angle allows you to specify an angle deviation between corresponding points on the edges of the quilt surface with the target surface faces. The default value is 90-degrees, which allows for any deviation.

Unigraphics begins with a simple approximation of the sheet and adds complexity until it falls within the specified tolerances. If the tolerance values are very small, Unigraphics creates a more complex quilt sheet. Hold the edge tolerance tight enough so that edge-to-edge deviations check with other sheets and are within modeling tolerances. You can loosen the interior tolerances so that Unigraphics has more flexibility and can create a simpler sheet than the original.

Show Checkpoint:

This displays the points calculated and used to approximate the sheet. This is a visual way to check the progress of the quilt while Unigraphics creates it. You can identify potential problem areas and modify them quicker using this aid.Using **Show Check Points** slows the process of creating the quilt but its benefits are greater than the time lost.

Check for Overlaps:

This forces Unigraphics to check the surrounding sheets/faces for overlaps at each projected point.

If you activate this option and use a vector direction as the projection method, it is important to use the correct positive or negative vector direction. If a negative direction is used and an overlap condition exists, Unigraphics uses the top sheet to project the points on until it ends and then the bottom sheet is used. If the direction vector is positive, the points project to the bottom sheet first.

Exercise: Creating a quilt surface with curves and along driver normals projection

1. Open **ff_46**.
2. Select **Insert | Free Form Feature | Quilt**.
3. Click **Along Driver Normals**.
4. Click **OK**.
5. Pick the two primary curves in the Graphics window.
6. Click **OK**.
7. Pick the two cross curves in the Graphics window.
8. Click **OK**.
9. Pick the target faces in the Graphics window.
10. Click **OK** to create the quilt surface.

Exercise: Creating a quilt surface with curves and along fixed vector projection

1. Open **ff_47**.
2. Select **Insert | Free Form Feature | Quilt**.
3. Activate **Mesh of Curves**.
4. Click **Along Fixed Vector**.
5. Type the appropriate values in the **Tolerances** text boxes.
6. Click **OK** to display the **Select Primary Curves** dialog.

7. Pick the two primary curves in the Graphics window.
8. Click **OK** to display the **Select Cross Curves** dialog.
9. Pick the two cross curves in the Graphics window.
10. Click **OK** to display the **Vector Constructor** dialog.
11. Define a vector using the dialog.
12. Click **OK** to display the **Class Selection** dialog.
13. Pick the target faces in the Graphics window.
14. Click **OK** to create the quilt surface.

Exercise: Creating a quilt surface with an existing B-surface and along driver normals

1. Open **ff_48**.
2. Select **Insert | Free Form Feature | Quilt**.
3. Activate **B-surface**.
4. Click **Along Driver Normals**.
5. Click **OK**.
6. Pick the B-surface in the Graphics window.
7. Pick the target faces in the Graphics window.
8. Click **OK** to create the quilt surface.

Exercise: Creating a quilt surface with an existing B-surface and along fixed vector

1. Open **ff_49**.
2. Select **Insert | Free Form Feature | Quilt**.
3. Activate **B-surface**.
4. Click **Along Fixed Vector**.
5. Click **OK** to display the **Select Driver B-surface** dialog.
6. Pick the B-surface in the Graphics window.
7. Define a vector using the dialog.
8. Click **OK**.
9. Pick the target faces in the Graphics window.
10. Click **OK** to create the quilt surface.

Exercise: Refitting a B-surface

1. **Open ff_50**.
2. Select **Insert | Free Form Feature | Quilt**.
3. Activate **Self-Refit**.
4. Click **OK**.
5. Pick the B-surface in the Graphics window.

Note: Quilting is a good option to use to simplify a complex sheet before using it to trim. It prevents the complexity of the trim sheet from passing to the target body.

Midsurface:

Insert | Free Form Feature | Midsurface

Create sheet bodies centered between two faces.

Unigraphics creates face pairs to define the inside and outside limits and then creates a midsurface feature trimmed to the surrounding geometry. This option is useful when creating geometry for thin wall analysis as well as rib and wall construction. The midsurface sheet is associated to the parent face pairs.

Auto face pairs runs a program that calculates as many face pairs in the part as possible. It may take several attempts to create all the face pairs needed to create the midsurface sheet. If face pairs already exist in the part, they are ignored by the system the next time **Auto Face Pairs** runs.

Generate face pairs are a more interactive way to create face pairs. The face pairs generate from a group of faces you select. Unigraphics generates the midsurface from these face pairs. If the desired result is a group of midsurfaces, but not midsurfaces of the entire solid, the **Generate Face Pairs** option is a good choice. Use this option to create any face pairs ignored by the **Auto Face Pairs** option.

After **Generate Face Pairs** is selected, Unigraphics prompts you for a "Non-Paired Side 1" or "Side 2." You may pick as many faces as you wish to generate face pairs from.After face selection, Unigraphics looks for opposing faces. If opposing faces are already selected, a midsurface is created between those faces. If an opposing face is not specified, Unigraphics finds the closest face opposing the selected faces and a midsurface is created between these face pairs.

Create Face Pair also creates a midsurface from selected faces. Unlike the **Generate Face Pairs** option, which creates multiple midsurface bodies, **Create Face Pair** only creates one midsurface using all of the faces selected. This is a good method to use if the desired result is one sheet between two sets of faces. If more than two opposing faces are selected, Unigraphics tries to approximate a sheet taking the average of all faces selected, sometimes producing an unwanted result. It is advisable to select Side 1 faces that are similar to Side 2 faces for a more predictable result.

After you select **Create Face Pair**, Unigraphics prompts you for Side 1 Faces. These faces highlight in cyan. After you finish selecting Side 1 faces, the Prompt asks for Side 2 Faces. These faces highlight in yellow.

When you select both side face sets, you need to specify the **Selection Type**.

> **Use Side Faces** creates a midsurface between only two faces. This option uses the faces from Side 1 and Side 2 to define the size and shape of the midsurface feature.

Use **Defining Faces** when multiple faces are selected for either side. When **Defining Faces** is selected, the prompt asks for **Selection Type**. You must specify the two faces to use to define the midsurface. Unigraphics creates the midsurface between these two faces but use all the faces selected to define the size and shape for trimming.

Midsurface operates just as the **Defining Faces** option but when satisfying the Selection Type prompt, you define the face to create the midsurface on.

There are two different midsurface features you can create using the Midsurface function: FACEPAIR_DEF or FACEPAIR_SEL. The type of feature created depends upon which method you used for creation.

The feature FACEPAIR_DEF is created when two opposing faces on the same target body are used as the defining faces for the midsurface. When you use only one face for defining the midsurface, the feature FACEPAIR_SEL is created.

Use the **Use Side Faces** and **Defining Faces** options to create FACEPAIR_DEF features and the **Midsurface** option to create FACEPAIR_SEL features.

Modifications in the model may not always update the user-defined Midsurface. If the trimming faces move, this usually carries over to the Midsurface. If thedefining faces are moved, the midsurface is not updated.

Offset method allows you to create a midsurface for a region of faces, and can be easier and more predictable to use than the **Auto Face Pairs** method. You can adjust the position of an offset midsurface so that it doesn't lie directly at the midline of the opposing faces.

Editing Midsurface features:

When creating midsurfaces, you can only run the Midsurface operation once. When you select **OK** or **Apply**, Unigraphics creates the midsurface bodies. You are unable to select the target solid or any subsequent Midsurface operations. If more midsurfaces are desired, the Midsurface feature can be edited using **Edit | Feature | Parameters**. This allows you to add or remove faces from the definition of the face pairs.

Exercise: Creating a Midsurface feature with auto face pairs

1. Open **ff_51**.
2. Select **Insert | Free Form Feature | Midsurface**.
3. Set **Method** to **Facepair**.
4. Click **Auto Face Pairs**.
5. Click **OK** to trim the midsurfaces to the bounds of the part.

Exercise: Creating a Midsurface feature with generate face pairs

1. Open **ff_52**.
2. Select **Insert | Free Form Feature | Midsurface**.
3. Set the **Method** to **Facepair**.
4. Click **Generate Face Pairs**.
5. Pick faces to define the midsurfaces.
6. Click **OK** to create the midsurfaces.
7. Click **OK** to trim the midsurfaces to the length of the selected faces.

Exercise: Creating a Midsurface feature with create face pair

1. Open **ff_53**.
2. Select **Insert | Free Form Feature | Midsurface**.
3. Set the **Method** to **Facepair**.
4. Click **Create Face Pair**.
5. Pick faces to be used to define one side of the midsurface.
6. Click **OK** to finish the selection of the Side 1 faces.
7. Pick faces to be used to define the opposite side of the midsurface.
8. Click **OK**. The faces you picked for Side 1 highlight in cyan, and the faces for Side 2 will highlight yellow in the Graphics window.
9. Click **OK** to create a midsurface.
10. Click **OK** to trim the midsurface to the bounding surfaces.

Exercise: Creating a Midsurface feature with offset

1. Open **ff_54**.
2. Select **Insert | Free Form Feature | Midsurface**.
3. Set the **Method** to **Offset**.
4. Pick the body to create a midsurface on.
5. Click **OK**.
6. Pick a seed face on the body. This face is used to measure the offset of the midsurface.
7. Select **Finish** to restore menu function.
8. Move the **Midsurface Position** slider bar to the desired percentage offset.
9. Click **OK** to create the midsurface.

Note: For all midsurface sheets the type of surface created is dependent upon the types of faces you select for the side faces. If a B-surface is selected as one of the side faces, a B-surface is created as the midsurface. The generated surface is an average of the two side faces at their parameter points.

Bridge (Surface):

 **Insert | Free Form Feature | Bridge**

Create a sheet that spans a gap between two existing sheets. The resultant sheet matches either curvature or tangency of the two existing sheets. This feature is similar to **Bridge Curve**. The bridge sheet is also associative to the reference sheets.

Side Faces:

 When using **Side faces**, Unigraphics prompts you to select a side face and places a point on the edge that it uses as the constraint. You can pick a side face for either side or both sides. **Side Face** creates a bridge sheet that is tangent continuous to all faces (primary and side faces).

Side String:

 Side String **allows you to select curves or edges to define the adjacent edges of the bridge sheet.**
When using the Continuity type **setting** Curvature**, the selected side sheets must be curvature continuous with the primary sheets. If they are not, an Error message appears.**
You must use the Tangent **Continuity type when choosing side strings. You can use one or two side strings.**
Continuity type:
You can define the continuity of the bridge sheet in two ways: Tangent **and** Curvature**.**

> **Tangent** creates a cubic blend that matches tangency at the defining sheet edges.
>
> **Curvature** creates a sheet that is tangent to the defining sheets and has the same rate of curvature where its edges meet with the defining sheets.

Drag:

It is possible to edit the bridge sheet during creation. Selecting **Drag** allows you to dynamically drag a row of poles from either end of the primary bridge sheet. While dragging does change the shape of the sheet, the continuity holds to what you chose. This option is only available when you select **Apply** in the **Bridge** dialog.
To drag the bridge sheet, click **Drag** and then pick the sheet at the end you want to modify. Pressing MB1, drag the sheet in the desired direction. **Drag** is not

available if you select **Side faces** or **Side strings**. **Reset** returns the bridge sheet to its original state before you executed **Drag**.

Exercise: Creating a bridge sheet body

1. Open **ff_55**.
2. Select **Insert | Free Form Feature | Bridge**.
3.  Click **Primary faces**.
4. Pick the face where the bridge sheet will start. Make sure to pick near the edge where you want the bridge sheet to begin.
5. Pick the face where the bridge sheet ends. Make sure to pick near the edge where you want the bridge sheet to end. Make sure the directional arrows point in the same direction.
6. If desired, use **Side faces** or the **Side String** options to provide more control over the shape of the bridge sheet.
7. Set the **Continuity type**.
8. Select **Apply**.
9. Select **Drag**, if desired.
10. Pick a point near an edge of the bridge sheet and drag.

Notes: When using **Side Face** or **Side String** to create a bridge, they must form a closed loop with the primary faces' edges.

Use the edges of the primary sheets to create a curvature continuous bridge curve that can then be used as the side string of a bridge sheet.

Thicken Sheet:

Insert | Form Feature | Thicken Sheet

Create a solid body from a sheet body by giving volume and mass to the sheet. The result is a feature (solid body) that is associative to the sheet.

Thickening is analogous to offsetting a sheet body a given distance normal at all points. You can select a single sheet body or multiple sheets sewn together. You can also select a sheet body with multiple interior trimmed edges (trimmed holes). This sheet thickens and leaves through holes in the solid body.

First and second offset are identical to the **Start Distance** and **End Distance** options for an extruded body, as they determine the start and end distances of the thickened sheet relative to the input sheet body. After selecting the sheet, a vector displaying the positive normal thickening direction is displayed centered on the sheet. Input negative values to thicken in the opposite direction. The thickness of the resulting body is determined by the difference between the **First** and **Second Offset** values.

Action allows you to incorporate a Boolean operation into the thicken sheet feature. If you set the **Action** to anything other than **Create**, the select **Target Solid Body** icon becomes available.

The default for **Action** is **Create**. If there is no solid in the part file, this is the only option available.

Exercise: Creating a thicken sheet body

1. Open **ff_56**.
2. Select **Insert | Form Feature | Thicken Sheet**.
3. Pick the sheet to thicken.
4. Type in values for the **First Offset** and **Second Offset**.
5. Set the **Action** to the appropriate Boolean method.
6. Pick a target solid body if you select a **Unite**, **Subtract**, or **Intersect**.
7. Select **OK**.

Trimmed Sheet:

Insert | Free Form Feature | Trimmed Sheet

Redefine the bounds of a sheet body.

You can use two methods to define the bounds: project curves or edges onto the sheet or intersect faces or datum planes with the target sheet.The sheet maintains its original parameters after trimming. It also becomes associative to the geometry used to define the trim.

Selection Steps:

Target Sheet Body

This picks the sheet body that you are trimming.

Projection Vector

The **Projection Vector** is the direction that the geometry, chosen to define the trim bounds, is projected onto the target sheet. This option is only active when you set **Projection Along** to **Datum Axis**.

Trim Boundary

This allows you to select the curves, edges, faces, and datum planes that define the trim edges. Curves and edges project onto the sheet, while faces and datum planes intersect with the sheet to produce the boundary. A Warning message appears when elements cannot project onto the sheet or fails to intersect the sheet.

Region

This works in conjunction with the **Regions will be Kept/Discarded** option (located lower in the dialog). Use this to select the regions of the trimmed sheet that you retain. While this is active, use the cursor in the Graphics window to click the area of the sheet to be kept/discarded.

Project along:

This determines the projection direction for the Trim Boundary curves.

Regions will be:

When you select a region, use this to specify whether the selected region is **Kept** or **Discarded**.

Tolerance:

Use this when the projected geometry imprints on the sheet. This controls the allowable tolerance of the trim edges projected on the target body.

Output exact geometry:

This produces a trimmed edge with zero deviation to a bounding sheet body.

Exercise: Creating a trimmed sheet

1. Open **ff_57**.
2. Select **Insert | Free Form Feature | Trimmed Sheet**.
3. Click **Target Sheet Body**.
4. Pick the sheet to trim in the Graphics window.
5. Click **Trim Boundary**.
6. Pick the trim boundary in the Graphics window.
7. Define the projection direction of the trim boundary geometry using the **Projection Along** list.

8.  Click **Region**.
9. Activate either **Kept** or **Discarded**.
10. If you select **Kept**, pick a point in the area you want to maintain. If you select **Discarded**, pick a point in the area you want to trim away.
11. Define the **Tolerance**.
12. Select **OK**.

Face Blend:

 **Insert | Feature Operation | Face Blend**

With this you can correct an edge blend failure.

Create a tangent blend that attaches to a solid or sheet body.

This blending feature uses two faces to control the blend whereas edge blend in the **Feature** dialog used edges. This fact allows the face blend to create a blend that eliminates multiple faces, edges, and vertices or blend faces on two different solids.

Selection Steps:

First Set

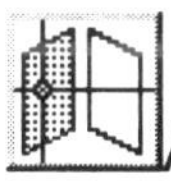Activate this icon to select the geometry defining the tangent face or faces for one side of the blend.

Second Set

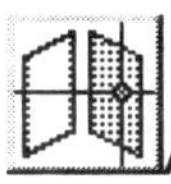Activate this icon to select the geometry defining the tangent face or faces for the other side of the blend.

Cliff Edges

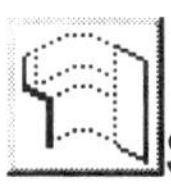Select this to define any faces where the blend is not tangent. If it is difficult to see where the cliff edges occur, pick all of the surrounding edges that might cliff. Unigraphics determines which edges to use and ignores the rest.

If **Include Tangent Faces** is active, you only need to select one cliff edge.

Unigraphics defines all edges tangent to it. The blend starts at the defined cliff edge and flows out tangent to the opposite face.

Tangency Control

This uses a selected tangent string to define the point of tangency of the blend. It is available on both spherical and conical face blends. You can define the radius of a spherical blend by the tangency string as well as where the blend becomes tangent. Set the **Radius Method** to **Tangency Controlled** to use this option.

For conical blends, the tangent string is used to compute the smallest offset on the face opposite the tangent string. A tangency string can only be chosen for one face set. It can only be used with the first face set if **Project on First Set** is active. You can set **Filter** to allow faces, edges, curves and datum planes to be selected as the tangency string.

Reverse Normals:

When selecting the first face of the first set and the first face of the second set, it is important to watch the NORMAL DIRECTION that displays in the Graphics window. The normal should be pointing towards the center of the blend. Use **Reverse Normal** to switch the direction.

Include Tangent Faces:

You only need to select one face when this option is active; Unigraphics define all the faces that are tangent to it.

End Tangent Overflow limits the use of additional tangent faces to one face beyond the end of a trimmed face set. If the faces, tangent to face set one, end before the faces, tangent to face set two, Unigraphics includes only the blending edges that result from the one tangent face of set two that is beyond the last tangent face of set one. This option also allows a blend to pass over a notch in the faces.

Attachment method defines the way Unigraphics trims the blend after it is created. All the Attachment Methods create parametric features.

Trim & Attach All trims all faces that are tangent to the face blend, including the face blend, and sews them into one sheet body or solid.

Trim All trims all faces tangent to the face blend, including the face blend, but does not perform the sew operation.

Trim Blend trims only the face blend to the limiting edges of the faces or to specify limiting planes.

Trim Blend Short trims the face blend to the edge that creates the shortest possible blend.

Trim Blend Long trims the face blend to the edge that creates the longest possible blend.

No Trim does not trim any faces.

Blend Type:

This controls the cross sectional shape of the blend. The four blend types are **Sphere**, **Conic**, **Disc**, and **Isoparameter.**

Sphere uses the same rolling ball process that defines an Edge Blend in the Feature Operation menu. The cross section is normal to the two face sets used to define it. There are three options in the spherical Radius Method: **Constant**, **Law Controlled**, and **Tangency Controlled**.

Conic creates a conic cross section controlled by the offset distances of the face sets and a Rho value.
The **First** and **Second Offsets** allow you to specify a distance using the text boxes. Their intersection is computed and a minimum distance represents a normal from the point of tangency to the intersection of the offsets. Both offsets can be defined as a constant to a law controlled value. The **Rho** value is used to define the type of conic to span the two points of tangency. The **Rho** value must be between .001 and .999. If the value entered is:

. 001 < Rho < .500 an ellipse is created
. 500 = Rho A parabola is created
. 500 < Rho < .999 A hyperbola is created.

The default for **Rho** is set to **Auto**; this allows Unigraphics to define Rho. This option setting lets Unigraphics establish a Rho that creates a circular blend if the offsets are equal otherwise the blend is elliptical. Rho can also be set to **Constant** or **Law Controlled**.

You must define a spine string for all conic face blends. This determines the normal plane of the conic cross sections.

Disc creates a law controlled variable radius blend between the two selected face sets. The difference between a disc and sphere blend is how the cross section of the blend is oriented as it follows the blend faces.

Imagine this blend being a zero thickness disc, with a set radius, swept along normal to the spine following the contour of the face sets. This allows a disc blend to blend faces with tighter curvature if a spine is chosen that prevents the cross sections from intersecting each other. A spine that is linear or with little curvature will maximize the blends potential size.

A disc blend requires a spine string to determine the blend's orientation. As the blend follows the selected blend faces, the cross section stays normal to the spine instead of the blend edge.

The only radius option available for a disc blend is **Define Law**. The **Cliff Edge** option is also active with disc blend.

Use **Isoparameter** when you are working with advanced geometric shapes (highly sculptured surfaces) where other types of blends may potentially fail.

Isoparameter is similar to the Disc blend where the sections through the blend are specified by reference geometry. In the case of the Disc blend, the reference geometry is the spine string. **Isoparameter** uses the isoperimetric curves of the first set faces. The Isoparameter Blend is fully parametric and associative to its reference geometry.

Radius Method is only available when creating a **Blend Type** Sphere.

Constant creates a circular cross section with a constant radius along the blend's length. The radius must be positive.

Law Controlled allows the radius to be defined by one of the Law methods found on the Law Subfunction dialog. You can create a variable or constant radius using the Law options. The Radius text box is inactive for this option.

Constant creates an unchanging radius blend defined by the law value.

Linear creates a variable blend that has a linear interpolation from the start point to the end point. On planar faces, this results in a straight-line effect on the tangency intersection of the blend.

Cubic creates a variable blend with a cubic interpolation from the start point to the end point. On planar faces this results in an S-shaped effect on the tangency intersection of the blend.

Values Along a Spine-Linear and **Values Along a Spine-Cubic** allow a variable blend to be created with different values entered at selected points along a specified spine. This option can have a linear or cubic interpolation along the length of the blend.

By Equation allows the variable radius to be defined along a specified spine using a Law function. An equation defined by a law function consists of two expressions: a function and a parameter. The function expression is defined in terms of the parameter with the parameter always having a value range from 0 to 1. For example:

Ft=10+4*t. This is the function in terms of **t**
t=1 this is the parameter for **Ft.**

The function value **Ft** varies as the parameter value moves from 0 to 1. The parameter **t** is mapped linearly along the length of the spine.

By Law Curve defines the variable radius using the Y values of a law curve. The Y values are measured from the law curve to the Absolute X-Axis as it is mapped out along the length of a specified spine curve. A base line may be specified for use in measuring the normal distance to the law curve defining the law function. This may be more convenient because of the location to the base line.

Tangency Controlled uses a tangent string projected onto one face set or the other to define the radius. You can only use one string to control the radius. The tangency string defines the attachment point of the blend to the selected face. You can use faces, edges, curves and datum places to define the string. Once you selected a tangency string, the **Project on First Set** option becomes active. If **Project on First Set** is checked, the tangency string projects normal to the first face set, otherwise it projects on the second face set. The **Radius** text field is inactive for this option.

Limit Start and End highlights if you set the Attachment Method to **Trim Blend** or **No Trim.** This determines the extent of the face blend by defining its endpoints.

Limit Start defines a start plane for the start of the blend by using the **Plane** dialog. When the plane is specified select **Accept** or **Reverse** for the direction of the blend. If the desired blend is to be between two planes, make sure the direction is pointing toward the limit end plane.

Limit End has the same steps as **Limit Start** but defines the end of the blend. Be sure the direction for **Limit End** is opposite of **Limit Start**. If it is going in the same direction as **Limit Start**, the blend is created outside the limit planes in the direction specified.

Help Point on the **Face Blend** dialog is available to help Unigraphics determine the faces to blend if there is more than one possibility. Once you click **Help Point**, the **Point Constructor** dialog displays so you can specify a point. For instance, if you selected a face for the first set that intersects the second face set in more than one place, Unigraphics tries to blend and trim both intersections. If this is not desirable, use a help point to define the area to blend.

Exercise: Creating a spherical face blend with a radius

1. Open **ff_58**.
2. Select **Insert | Feature Operation | Face Blend**.
3. 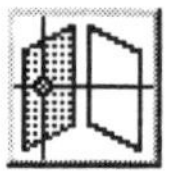 Click **First Set**.
4. Pick the first set of faces. The normal must be pointing toward the center of the face blend. If it is not, click **Reverse Normal**.
5.  Click **Second Set.**
6. Pick the second set of faces. The normal must be pointing toward the center of the face blend. If it is not, click **Reverse Normal**.
7. 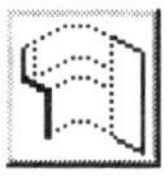 Click **Cliff Edges** if required.
8. Select **Sphere** for the **Blend Type** and choose an **Attachment Method**.
9. Type in a value in the **Radius** text box or select **Law Controlled** and define a law function.
10. Click **OK** twice to create the face blend.

Exercise: Creating a spherical face blend with tangency control

1. Open **ff_59**.
2. Select **Insert | Feature Operation | Face Blend**.
3. 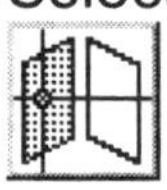 Click **First Set**.
4. Pick the first set of faces. The normal must be pointing toward the center of the face blend. If it is not, click **Reverse Normal**.
5.  Click **Second Set**.
6. Pick the second set of faces. The normal must be pointing toward the center of the face blend. If it is not, click **Reverse Normal**.
7. 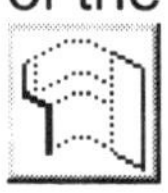 Click **Cliff Edges** if required.
8. Select **Sphere** for the **Blend Type** and choose an **Attachment Method**.
9. Change the **Radius Method** to **Tangency Controlled**.
10.  Click **Tangency Control**.
11. Pick a set of curves from the Graphics window.
12. Click **OK** twice to create the face blend.

Exercise: Creating a conical face blend

1. Open **ff_60**.
2. Select **Insert | Feature Operation | Face Blend**.
3.  Click **First Set**.
4. Pick the first set of faces. The normal must be pointing toward the center of the face blend. If it is not, click **Reverse Normal**.
5. 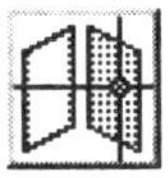 Click **Second Set.**
6. Pick the second set of faces. The normal must be pointing toward the center of the face blend. If it is not, click **Reverse Normal**.
7. Select **Conic** for the **Blend Type** and choose an **Attachment Method**.
8. Type in a value for **First Offset**, **Second Offset**, and **Rho** or choose **Law Controlled**.
9. Click **Define Spine String**.
10. Pick a set of curve strings from the Graphics window.
11. Click **OK**.
12. Click **OK** twice to create the face blend.

Exercise: Creating a disk or isoperimetric face blend

1. Open **ff_61**.
2. Select **Insert | Feature Operation | Face Blend**.
3. 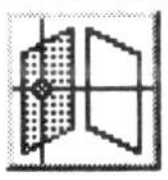 Click **First Set**.
4. Pick the first set of faces. The normal must be pointing toward the center of the face blend. If it is not, click **Reverse Normal**.
5.  Click **Second Set.**
6. Pick the second set of faces. The normal must be pointing toward the center of the face blend. If it is not, click **Reverse Normal**.
7. Select **Disc** or **Isoparameter** for the **Blend Type** and choose an **Attachment Method**.
8. Click **Define Law**.
9. Define the radius using a Law function.
10. Define a spine string.
11. Click **OK** to return to the **Face Blend** dialog.
12.  Click **Cliff Edges** if required.
13. Click **OK** twice to create the face blend.

Notes: If for some reason the Face Sets need to be modified prior to creating the blend, simply reselect the appropriate face set icon, deselect the face/body to be replaced, and select a new face/body to be used for the face blend.

When creating a variable radius blend using **By Equation**, be sure the equation does not go to infinity. This may result in your session locking up. For example, the equation Ft= (tan(t*90)) results in an error since tan90=infinity.

Soft Blend:

Insert | Feature Operations | Soft Blend

Create a face blend where you control tangency on both faces with tangency control curves. This allows you to hold a particular tangency point but still have flexibility of the cross sectional shape.

Create a face blend where you skew the blend towards Face Set One or Face Set Two. This creates an aesthetically pleasing noncircular (conical) cross-sectional face. Modifying the shape of these blends can create designs with lower weights or stress resistant properties.

Selection Steps:

First Set

Pick the geometry to define the tangent face or faces for one side of the blend.

Second Set

Pick the geometry to define the tangent face or faces for the other side of the blend.

First Tangency Curve

The **First Tangency Curve** is a string of curves lying on the tangent face or faces defined in the **First Set**. The first tangency curve defines the attachment edge on one side of the blend.

Second Tangency Curve

The **Second Tangency Curve** is a string of curves lying on the tangent face or faces defined in the **Second Set**. The second tangency curve defines the attachment edge on the other side of the blend. The tangency curves define the size of the resultant soft blend

Smoothness:

Match Tangency creates an elliptical blend tangent to the two faces specified. **Rho** and **Skew** options are unavailable for this method.

Match Curvature creates a blend both tangent and curvature continuous at the point of attachment. When using this option, the shape and type of conic blend created is defined by the values entered for **Rho** and **Skew**.

The **Rho** value determines the type of conic you create. If .001 < Rho <.500, the conic is elliptical; if Rho =.500 the conic is parabolic; if.500 < Rho < .999, the conic is hyperbolic. A Rho value near zero creates a flatter shaped blend while a value approaching 1.0 creates a more peaked blend.

Skew is a percentage value (0.01 to 0.99) that determines how much the blend is pulled towards the second attachment face. A value near zero puts the blend's peak nearer the first attachment face. A 0.5 value represents a neutral skew influenced equally between the two faces. This is where the default conic is created. A skew value approaching 1.0 moves the blends peak toward the second attachment face.

The figure shows three different soft blends. Each blend has the same **Rho** value but different skew values.

Exercise: Creating a soft blend

1. Open **ff_63**.
2. Select **Insert** | **Feature Operations** | **Soft Blend**.
3. 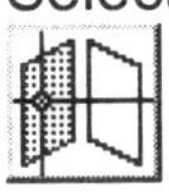 Click **First Set**.
4. Pick the first set of faces in the Graphics window and verify the normal is pointing towards the center of the blend. If it is not, click **Reverse Normal**.
5. 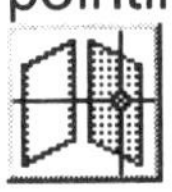 Click **Second Set.**
6. Pick the second set of faces in the Graphics window and verify the normal is pointing towards the center of the blend. If it is not, click **Reverse Normal**.
7.  Click **First Tangency Curve**.
8.  Click **Second Tangency Curve**.
9. Set the **Attachment Method** and the **Smoothness**.
10. Type in the **Rho** and **Skew** values if you checked **Match Curvature**.

11. Click **Define Spine String** to display the Spine String dialog.
12. Pick the Spine String in the Graphics window.
13. Click **OK** to return to the **Soft Blend** dialog.
14. Type in the **Tolerance** value if needed.
15. Click **OK** twice to create the blend.

Editing Freeform Features:

You can edit any free form feature created using parametric methods by selecting **Edit | Feature | Parameters**. Most free form body types have their own specific parameter-editing dialog. Each parameter option depends on the geometry used to create the body. Many parameter selections within the different dialogs are common from type to type. You cannot edit unparameterized sheets using this method. If you select an unparameterized feature, the following error message displays on the Status line: "Unparameterized Feature Is Invalid For This Operation."

When you activate this option, feature edits you apply do not immediately apply. When the edits are completed, click **Update** to perform the revision to the free form features. It is not necessary to click **Update**. The model automatically updates when you select any menu bar command.

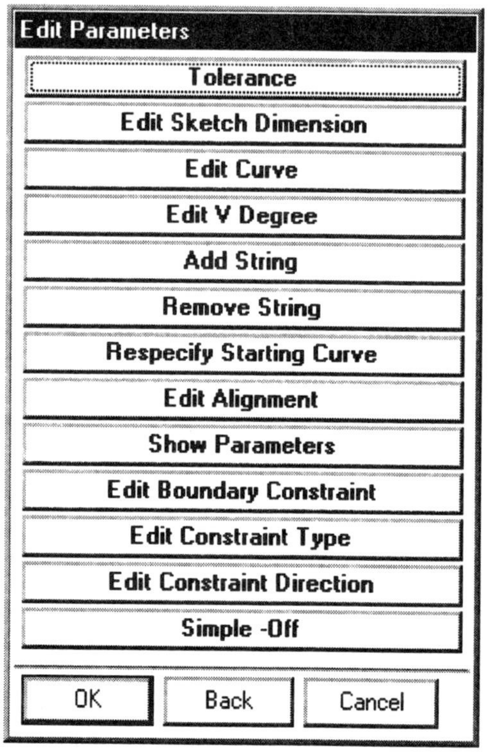

Editing through Curves Features

Edit | Feature | Parameters

Tolerance:

This edits the distance and angle tolerance for a free form feature. This modifies how closely the free form feature follows its original curves.

Edit Sketch Dimensions:

This displays on the dialog if you used sketches to create the section strings. You can edit the expression value of dimensional constraints for any sketch used in the Through Curves sheet creation. After selecting this option, pick a sketch from the Graphics window. A dialog displays prompting you to either accept or reject your selection.

After you accept a sketch, the **Edit Sketch Dimensions** dialog displays. You can change dimensional values, get an alternate solution to the sketch, or reattach the sketch to a new face or datum plane.

Edit Curve uses the curve editing options found under **Edit | Curve.** By editing the original curves of the free form feature, the size and shape of the feature can change. Once you select **Edit Curve,** the **Edit Curve** dialog displays.

Edit V degree changes the V degree of the selected feature if it is a multi-patch sheet. The maximum V degree is one less than the number of rows or section strings used.

Add String adds additional section strings to a free form sheet. After you click **Add String**, pick the existing section string you want to add the new string after. You have the same options as when creating a Through Curves feature. If you pick the first string, you can add the new string after or before it.

Remove String removes a section string from a free form feature. The minimum number of section strings a multi patch free form sheet requires is the degree plus one.

Respecify Starting Curve is used to switch the direction vector of a section string and or if there is a twist in a free form feature.

Edit Alignment changes the alignment method of a free form feature. The alignment method controls how the free form feature is mapped to the section strings. The only alignment methods available are **Parameter, Arclength,** and **By Points.** If you did not specify **By Points** during creation, this option is not available.

Show Parameters displays the section string names and direction vectors of a free form feature.

Edit Boundary Constraint adds a tangency or curvature constraint to the Through Curves feature's first or last section string.

Edit Constraint Type changes the type of constraint on the first and last section string. You can also use this option to remove the constraint from either string.

Edit Constraint Direction changes the constraint direction you set in creating the through curve sheet.

Simple on off changes the sheet from a simple construction to a normal sheet and vice versa. By using the **Simple** option, you can create a surface with a minimum number of poles and patches. The degree, poles and patches of a simple surface relate directly to the degree, poles and knots of the defining splines.

Exercise: Editing a through curve

1. Open **ff_63**.
2. Select **Edit | Feature | Parameters**.
3. Select **Through Curves**.
4. Click **OK**.
5. Click the appropriate option and perform the edit.
6. Click **OK**.
7. Click **OK** to perform the edit.

Notes: The **Edit Sketch Dimension** option does not work on pre-V13 sketches.

The **Add String** option rearranges the order of the section strings. It does not transform the strings to a new location. The section string added must be located in the proper location prior to being added to the free form feature.

Edit Boundary Constraint is a valuable tool for improving the quality of existing data.

Editing through Curve mesh:

 **Edit | Feature | Parameters**

Edit a feature created using Through Curve Mesh.

Many of the options for Through Curve Mesh features are the same as for Through Curves.

Tolerance edits the intersection, distance and angle tolerance for a freeform feature. This modifies how closely the free form feature follows its original curves.

Edit Boundary Constraints will add tangency or curvature constraints to your first and last primary and cross strings.

Special Types changes the construction method of the Through Curve Mesh Surface.

> **Off** resets the sheet to the normal construction method. This is usually the most complex type of surface.
>
> **Spline Pts** uses the definition points of the defining strings to create the sheet body instead of the strings themselves.
>
> Use **Simple** to create a surface with a minimum number of poles and patches. The degree, poles and patches of a simple surface relate directly to the degree, poles and knots of the defining splines.

Exercise: Editing a through curve mesh

1. Open **ff_64**.
2. Select **Edit | Feature | Parameters**.
3. Select **Through Curve Mesh**.
4. Click **OK**.
5. Click the appropriate option and perform the edit.
6. Click **OK**.
7. Click **OK** to perform the edit.

Editing Swept Features:

 **Edit | Feature | Parameters**

Edit Blend:

Use this to change the transition method, Linear or Cubic, between section strings. This option is only available when you have multiple section strings.

Display Parameters:

This displays the guide string names, section string names, and direction vectors of the section strings.

Exercise: Editing a swept feature

1. Open **ff_65**.
2. Select **Edit | Feature | Parameters**.
3. Select **Swept** feature from the list.
4. Click **OK**.
5. Click the appropriate option and perform the edit.
6. Click **OK**.
7. Click **OK** to perform the edit.

Editing Extension Sheets:

When you select an extension sheet body from the **Feature Selection** dialog, an **Edit Extension** dialog displays. The type of extension dictates the particular dialog that displays.

All the edit extension sheet body dialogs allow for tolerance substitution. It is impossible to parametrically edit a law controlled extension sheet since they are created as unparameterized features.

Editing Offset Surfaces:

 **Edit | Feature | Parameters**

Edit the offset value or change the target Face of an offset face.

Edit the offset values of a variable offset sheet. You can also edit the location of the variable offset.

Offset distance 1, 2, 3 and 4 displays the offset distance for each corner of the variable offset sheet. You can change the values of the offset and even input negative numbers. The corresponding corner displays a numbered marker in the Graphics window.

U and V parameters correspond to the U and V parameter percentage at each of the four corners. The default values are 0 and 1 with 0 being 0% and 1 being 100%. You can change these to any value between 0 and 1. The offset distance corresponding to the U and V parameter offsets at the new location.

Exercise: Editing the offset of a fixed offset sheet

1. Open **ff_66**.

2. Select **Edit | Feature | Parameters**.
3. Select **Offset**.
4. Click **OK**.
5. Click **Edit Creation Parameters**.
6. Enter new values for the **Offset Distance** and/or **Tolerance**.
7. Click **OK**.
8. Click **OK**.
9. Click **OK** to edit the offset sheet.

Exercise: Changing the target face

1. Open **ff_67**.
2. Select **Edit | Feature | Parameters**.
3. Select **Offset**.
4. Click **OK**.
5. Click **Change Face**.
6. Pick a new target face from the Graphics window.
7. Click **OK**.
8. Click **OK**.
9. Click **OK** to edit the offset sheet.

Exercise: Changing a variable offset sheet

1. Open **ff_68**.
2. Select **Edit | Feature | Parameters**.
3. Select **Var Offset Sheet**.
4. Click **OK**.
5. Change the Offset Distances, U Parameters, or V Parameters.
6. Click **OK**.
7. Click **OK** to edit the offset sheet.

Editing Bounded Planes:

 **Edit | Feature | Parameters**

Edit defining curves and sketch dimensions.

Add or remove geometry from the defining strings. You must remove curves before adding new curves.

Exercise: Redefine the surface by removing and adding strings

1. Open **ff_69**.
2. Select **Edit | Feature | Parameters**.
3. Select **Bounded Plane**.
4. Click **OK**.

5. Click **Remove String**.
6. Pick the string that defines the bounded plane.
7. Click **OK**.
8. Click **Add String**.
9. Pick the new geometry from the Graphics window.
10. Click **OK** after picking all the new geometry.
11. Click **OK**.
12. Click **OK** to edit the bounded plane.

Editing Ruled Features:

 **Edit | Feature | Parameters**

Edit the parameters of a ruled body. The options available are the same as found for Through Curves.

Add/Remove String:

Although these add and remove strings from the ruled body, you can only have two strings. If you add a string, you must remove one of the original strings.

Exercise: Editing a ruled body

1. Open **ff_70**.
2. Select **Edit** | **Feature** | **Parameters**.
3. Select **Ruled**.
4. Click **OK**.
5. Edit the desired parameter.
6. Click **OK**.
7. Click **OK** to edit the ruled body.

Sew:

 **Insert | Feature Operation | Sew**

Join two or more sheet bodies into a single sheet body. Join two or more sheet bodies that enclose a volume into a solid body. Join two or more sheet bodies together to create tool solids. Join two or more sheet bodies together to create a contoured area that might not be possible to create using solid features. Join two or more sheet bodies together to be thickened to create a solid body. Join two solid bodies together by sewing faces of the solids.

Sew Input Type:

Sheet joins together sheet bodies. **Solid** joins faces of two solids together.

Target Sheet is a sheet body the Tool Sheets are sewn to.

Tool Sheet is the sheet body(s) to be sewn to the target sheet. You can drag a rectangle around the sheets for easier selection.

Use **Target Faces** to pick the face(s) on a solid to sew together with coincident face(s) of another solid.

Tool faces picks the face(s) of a solid coincident with the target face(s).

Output Multiple Sheets will allow the creation of more than one sewn sheet. This option is only available for Sew Sheets.

Sew All Instances sews an entire instance array if the selected tool sheet is a member of an array. If **Sew All Instances** is inactive, only the selected sheet(s) sew together.

Sew Tolerance lets you control the deviation allowed between the sewn edges. The default sew tolerance is the modeling distance tolerance. The operation fails if the sewn edge or face distance is greater than the entered sew tolerance.

It is important to control the amount of deviation allowed for a sewn sheet body. If you use the **Sew** command to trim a solid body or to make a tool solid, the tolerance should not exceed the modeling distance tolerance. If you use the feature to re-create reference geometry or to create a solid for calculating volume and mass estimates, a looser tolerance may be acceptable. If the sew operation fails because of an excessively large edge deviation, then modify the input sheets before loosening the tolerance.

When sewing sheets together to create a solid body there must be no edge-to-edge deviations greater than the specified tolerance, otherwise you may create a sewn sheet.

Search Common Faces finds faces between solids that are common to each other and highlights them.

Target Area displays the total area of the target faces.

Tool Area displays the total area of the selected tool faces.

Exercise: Sewing sheets

1. Open **ff_71**.
2. Select **Insert | Feature Operation | Sew**.
3. Activate **Sheet**.

4. 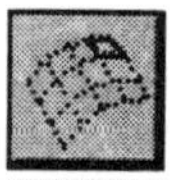 Click **Target sheet** and pick a sheet body.
5.  Click **Tool sheets** and pick the sheet body(s).
7. Enter a value for **Sew Tolerance**.
8. Click **OK**.

After the sheet body is created all of the free edges highlight in the system color. A message window appears if the operation was not successful for some of the sheets.

Exercise: Sew solids

1. Open **ff_72**.
2. Select **Insert | Feature Operation | Sew**.
3. Activate **Solid**.
4.  Click **Target faces** and pick the face(s) on the target solid.
5.  Click **Tool faces** and pick the face(s) on the toll solid.
7. Enter a value for the **Sew Tolerance**.
8. Click **OK**.

Notes: Although sewn sheets are treated as one sheet or solid body, the gaps on the edges between the sheets are not actually "matched" to each other. These gaps are all smaller than the Distance Tolerance specified (otherwise the "Sheets are Disjoint" message would be displayed), but there are still gaps. Therefore, it is extremely important that careful consideration is given to specifying the appropriate tolerance so downstream applications, such as Manufacturing, can successfully use the sewn sheets.

The Sew feature remains associative to the input sheets and updates when modifications are made to either the input sheets or their creation curves. This feature can be suppressed, unsuppressed, or deleted.
If the deviation between the edges is greater than the tolerance being used, the edges can be matched using the **Edit | Free Form Feature | Change Edge** option.

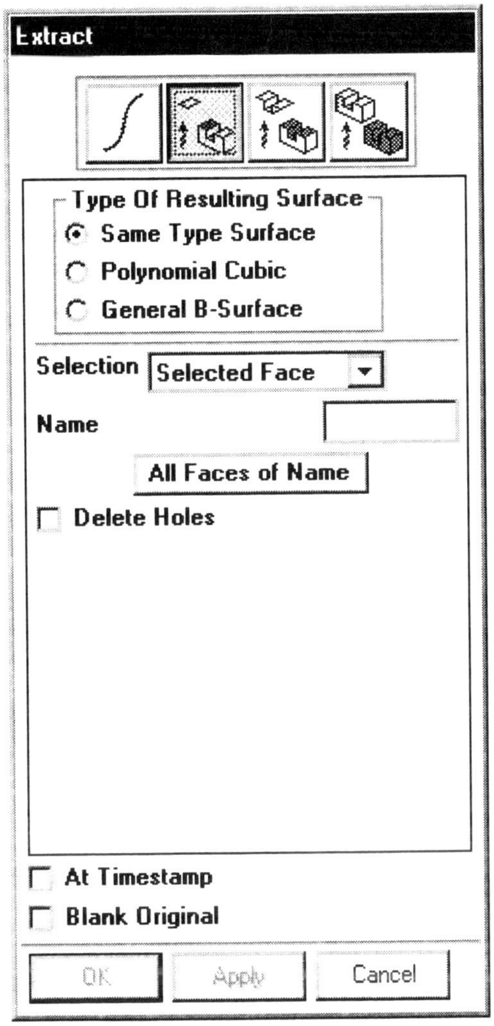

Extract:

Extract creates an associative copy of geometry. Modifications made to the original geometry can also appear in the extracted geometry, if desired.

You can extract four types of geometry: **Curve**, **Face**, **Region** and **Body**.

Curve allows you to extract curves from a solid object while maintaining associativity. You can also extract edges, wire frame curves and sketch curves. Deselect curves and edges by pressing SHIFT and picking the curves. The extracted curves form a feature called EXTRACTED_CURVE.

This is different from **Insert** | **Curve Operation** | **Extract** in that it does not create an associative curve, nor does it create a feature.

You cannot edit an EXTRACTED_CURVE feature using the **Edit** | **Curve** options unless you first break its associative link.

Face allows you to extract an associative copy of a face. The face converts to its underlying B-surface type sheet body. The resulting sheet body is coincident to and linked to the original geometry. This option is useful for generating trimming surfaces and surfaces to export. The extracted face forms a feature called EXTRACT_FACE.

Region creates a sheet body from several contiguous solid faces. When you extract a region, Unigraphics creates an EXTRACT_REGION feature.

Body creates an associative copy of an entire solid body.Changes made to an extracted body are not reflected in the original solid; however, changes made to the original body may or may not be passed on to the copy. This is determined by the**At Timestamp** option. When you extract a body, Unigraphics creates an EXTRACT_BODY feature.

At timestamp determines how modifications to the original solid are passed on to the extracted copy.

If **At Timestamp** is active, the extracted feature does include subsequent modifications made to the original body. The extracted face or body can be reordered before other feature in the original body.

If **At Timestamp** is inactive, the extracted feature always includes modifications made to the original body. It automatically reorders the extracted feature to become the last feature in the model.

Blank original geometry when you create the extracted feature. This can make it easier to select the extracted feature (and not the original) for subsequent operations.

Note: Use **Insert | Curve Operation | Extract** to extract a curve representing an edge that will later be eliminated by a blend. You can reference the original edge after you eliminate it.

Extracting Faces:

 **Insert | Form Feature | Extract**

Three different face types can be created by the Extract operation.

> **Same Type Surface** converts selected faces into sheets and maintains the original sheet's surface type.
>
> **Polynomial Cubic** converts selected face to a B-surface that is approximated using the distance and angle tolerances defined using **Preferences | Modeling**.

General B-Surface converts the face to a B-surface using an exact representation of the original face. The extracted face created is more complex thana polynomial cubic sheet and is more difficult to transfer to other systems.

Selection helps pick faces by setting the selection type to **Selected Face**, **Adjacent to Face**, or **All in Body**. You can also select faces with name attributes by typing it in the **Name** text box. If there are multiple faces with the same name, use the **All Faces of Name** button.

Delete Holes automatically close all holes in the extracted face. **Delete Holes** is especially useful when extracting faces for trimming purposes.

Exercise: Extracting faces

1. Open **ff_74**.
2. Select **Insert | Form Feature | Extract**.
3.  Click **Face**.
4. Select **Type Of Resulting Surface**.
5. Activate **At Timestamp** if needed.
6. Activate **Blank Original** if needed.
7. Click **OK** to create the extracted face(s).

Extracting a Region:

 **Insert | Form Feature | Extract**

This creates an associative sheet body linked to a group of contiguous faces on a body.

Seed Face:

 This face defines a face in the region to extract. The extracting region propagates out from the selected seed face to include all adjacent faces inside the boundary.

Boundary Faces:

 These faces define the boundary of the region to extract by surrounding the region including the seed face.

Tangential Edge Angle:

Faces are included in a region in an outward direction from a seed face. When the normals of two adjacent faces exceed the tangential edge angle tolerance, the outward face is not included in the face region.

Traverse Interior Edges:

This includes faces whose edges form a closed loop on the interior of a selected face.

Delete all Openings:

This removes all openings in the extracted region. An opening is a closed interior loop of edges with faces on only one side of the edge.

Preview Region:

This button allows you to preview the extracted region before executing the operation.

Exercise: Extracting a region

1. Open **ff_75**.
2. Select **Insert | Form Feature | Extract**.
3. Click **Region**.
4. Click **Boundary Faces**.
5. Activate **Tangential Edge Angle** and enter an **Angle Tol (Deg)** value if needed.
6. Activate **Traverse Interior Edges** if needed.
7. Activate **Delete All Openings** if needed.
8. Activate **At Timestamp** if needed.
9. If desired, click **Preview Region** to view the region to be extracted.
10. Activate **Blank Original** if needed.
11. Click **OK** or **Apply** to create the extracted region.

Extracting a Body:

Insert | Form Feature | Extract

This creates an associative copy of a complete solid body.

Exercise: Extracting a body

1. Open **ff_76**.

2. Select **Insert | Form Feature | Extract**.
3. Click **Body**.
4. Pick the solid body or bodies to extract.
5. Activate **At Timestamp** if needed.
6. Activate **Blank Original** if desired.
7. Click **OK** to create the extracted body.

Notes: You can use this command to create an associative copy of a solid for tool design. You can scale and subtract the extracted body or unite it with a core or cavity model. The scale factor does not affect the original model. However, changes to the original model are reflected in the copy.

Use **Extract Body** to create tool solids. When the tool solids are later used for Boolean operations, the original solid remains visible while the extracted tool solid "disappears" as a result of the Boolean feature created.

Use **Extract Body** with the **At Timestamp** option active to create copies of a model showing various stages of a manufacturing process.

Patch Body:

Insert | Feature Operation | Patch

Modify a solid/sheet body by replacing or adding to the original faces withthose of a second sheet body.

Sew sheets together where the Sew operation may not work. The tool sheet does not need to be exact because the operation trims it back to the intersection.

Trim a solid body, leaving a feature you can edit (similar to**Trim and Attach All** in the Face Blend feature).

Add material to a solid body, leaving a local region that defines a particular contour. You can edit this area without affecting the entire face as shown in the following figure.

Selection Steps:

Use this option to pick the target body. The target body is the body to modify.

Use **Tool Sheet** to pick the sheet body whose properties transfer to the solid.

Tool Face is an optional step allowing you to pick a single face from a multiple face sheet body.

Create a Hole Patch:

This option patches a closed set of sheets onto a target solid and then removes the material inside the sheets to create a hole. This ability is useful when creating a hole with complex geometry.

Note: You can also achieve the same results by using the Trim Body feature.

Reverse Normal Direction:

Use this to reverse the direction of the normal of the tool sheet. The vector should point towards the material to add.

Exercise: Creating a patch body

1. Open **ff_77**.
2. Select **Insert | Feature Operation | Patch**.
3. Click **Target Body.**
4. Click **Tool Sheet.**
5. If the Tool Sheet has multiple faces and you only want to use one of them, click **Tool Face**.
6. If you selected **Tool Face**, pick the desired face in the Graphics window.
7. Verify the correct direction for the surface normal, and select **Reverse Removal Direction** if needed.
8. Click **OK** or **Apply** to create the patch.

Exercise: Creating a hole with patch body

1. Open **ff_78**.
2. Select **Insert | Feature Operation | Patch**.
3. Select **Create Hole Patch**.
4. Click **Target Body**.
5. Click **Tool Sheet**.
6. Click **OK** or **Apply**.

Notes: You can modify the base solid/sheet body by adding or subtracting material or by replacing faces.

When using the Patch with sheets, the result is similar to trimming a base sheet to a patch sheet and sewing them together.
When creating a hole patch, you must select **Create Hole Patch** before selecting the tool sheet.

Summary:

In this chapter, you learned how to:

- Recognize the Continuity, Degree, and Stiffness of a Spline.
- Create Bounded Planes Using a Closed Planar Profile of Curves.
- Use the Ruled Surface Feature to Create a Linear Transition between Section Strings.
- Use the through Curve Feature to Create a Smooth Transition between Section Strings.
- Use the through Mesh Feature to Create Surfaces Using Section and Cross-Strings.
- Create Bodies Using Rows and Columns of Points.
- Create Bodies Using Points from a Data File.
- Use the Common Section Options.
- Create Different Types of Sections.
- Create the Different Types of Extension Sheets.
- Offset Sheet Bodies using Constant, Variable, or Multiple Sheet Bodies.
- Quilt Surfaces Together Using Existing Sheet Bodies.
- Create Bridge Surfaces between Existing Sheets.
- Thicken Sheets to Create a Solid Body.
- Use **Free Form Feature | Trimmed Sheet** to Limit the Boundaries of Sheet Bodies.
- Create Face Blends Using Sets of Faces.
- Create Soft Blends Using Sets of Faces and Tangency Control Strings.
- Edit any Type of Free Form Feature Created Parametrically.
- Create Bodies Using a Point Cloud.
- Use the Swept Feature to Create Bodies Using Guide Strings.
- Extract Associative Copies of Faces and Face Regions.
- Extract Associative Copies of Bodies.
- Sew Sheet Bodies Together.
- Patch Bodies.